Contents under Pressure

One man's triumph over Chiari syndrome

Raphael D'Alonzo

Edited by: Diane E. Baumer, A.D.N., B.A.

Table of Contents

Preface

As the Founder and Executive Director of a non-profit dedicated to helping Chiari patients (Conquer Chiari, www.conquerchiari.org) I have had the opportunity to communicate with hundreds, if not thousands, of patients, family members, doctors, surgeons, and others interested in or touched by Chiari. As I've listened, I've heard stories of heart wrenching grief, of lives turned upside down, and of dreams destroyed; but I've also heard stories that bear testament to the power of the human spirit, the ability to struggle against what appear to be insurmountable odds, of families coming together and rallying around a loved one, and of professionals who have dedicated their lives to help those affected.

Through it all, I've come to realize that while on one hand everyone's Chiari story is unique, there is also a commonality which results in the sense of a shared experience for most Chiari patients. The differences in the Chiari experience are many: any two people are likely to experience different symptoms; one person may be diagnosed early, while another toils for years in the medical system; one person may come through mostly unscathed, while another's life is irreparably changed; one person may find that their doctors and loved ones are very understanding and supportive, while another may face doubt and denial even from their family.

The paths traveled on the Chiari journey are nearly infinite in their variety, yet they all share certain core elements in common. Everyone with Chiari is confronted with their diagnosis, whether it was a shock or a long-sought-for validation. Everyone must go through the process of trying to understand what they are facing and telling their loved ones there is something wrong. Everyone must look into the darkness and fear of not knowing what will happen. Everyone endures the intellectual and logistical challenges of finding a doctor and deciding what to do. Everyone rides the emotional rollercoaster as the mind tries to come to grips with what is happening. And in the end, everyone must adapt to their new life with Chiari. Whether it landed a glancing blow, or left deep, permanent wounds, everyone must adjust and move on.

The Chiari stories I've heard reflect this odd combination of uniqueness and commonality; each person different, yet somehow the same. Ray's incredible story reflects this as well. Like many, Ray was plagued by a number of symptoms which did not yield to an initial exam, and as such were written off as due to stress. Research has shown that the vast majority of Chiari patients suffer from five or more symptoms. At the same time, a study by Dr. Thomas Milhorat found that more than half of Chiari patients had been told at one time they were suffering from a mental problem. Interestingly, Ray did not experience the single most common Chiari symptom, namely a headache in the back of the head, brought on by straining, coughing, sneezing, exercise, etc.

As Ray began to deteriorate, with no explanation for what was occurring, he became deeply depressed, and any new symptoms which occurred were ascribed as being psychological in origin. Because of this, it took a year and a half until Ray was properly diagnosed, which believe it or not, is actually quicker than many. While not definitive, research has indicated that the average time for a Chiari patient to be properly diagnosed is likely more than five years.

Ray's bout with depression is also, sadly, all too common. Rates of depression, and other mood and cognitive problems, appear to be fairly high in the Chiari community, however it is not known why. Is there a direct link between Chiari itself and depression? Or, does being told nothing is wrong with you when you know something is, result in problems? Today, we don't know the answer to this question as the psychological effects of Chiari, both on mood and cognitive function, have not been studied.

Like most people with serious symptoms, Ray underwent surgery in an attempt to find relief. Chiari surgery is very traumatic, with part of the skull being removed, and for many, the covering of the brain being opened and expanded with a patch. Afterwards, Ray faced the long road to recovery, and again like many, he faced it with little guidance on what would work and what wouldn't.

Chiari and its effects can take a long time to recover from, and unfortunately, some people never do. Upwards of 50% of people live with permanent symptoms - such as pain - and deficits, and many are left disabled, depressed, and socially withdrawn.

What makes Ray's story incredible, and different from so many of the stories I've heard, is the perseverance he showed in his recovery. Not willing to give up the life he lived before, and despite not getting better quickly, Ray chose to focus on a lofty goal; to run another marathon.

Some people are hardly affected by Chiari, and after surgery they return to the life they were leading. I've sometimes wondered if they realize how lucky they've been; how close they came to a life altering event. Others, way too many in fact, have their lives blown apart by Chiari. They never recover fully; they lose jobs; they lose their family; and their dreams and hopes lie shattered at their feet.

Then there are a few, like Ray, who are dragged down to the depths of despair, but through a combination of luck, faith, and strength of will, are still able to emerge triumphant. I believe, through my own experience with Chiari and from listening to others, that the key to success for people to reclaim their lives, like Ray did, is to just keep pushing.

Through the inevitable ups and downs, triumphs and defeats, you must keep going. It is not easy, and obviously people must deal with real limitations. But at the same time, I truly believe that each person with Chiari should never give up hope, should never stop trying, and should live their life to the fullest. I sincerely hope that Ray's story of triumph will inspire others to reach out and take back what they have lost.

Rick Labuda

Chapter 1
Like Saul's awakening

I began to shake uncontrollably as if I had been immersed in ice water. Voices were calling out. "Hold him down. Hold him down. I need someone else to hold his legs. Get more warm blankets." I could hear but couldn't respond or open my eyes. I could also feel the warm blankets and the many hands restraining my frigid quivering body. Someone said, "This guy is really bucking." Someone else told about his experience with another patient having an allergic reaction to the anesthesia. A third person holding me down said, "That's nothing. Last week we had a guy in here that started laughing as he was coming out of it. He laughed so hard, we all began to laugh with him." Then I stopped shaking and opened my eyes. Someone told me, "You're all right. The surgery went fine. You were just having a reaction to the anesthesia." I was then asked how much pain I was in on a scale from 1 to 10 with 10 being the worst pain I could imagine. For some reason, I took the question very seriously. I thought, "This is the absolute worse pain I have ever experienced but I guess it could be worse. I guess someone could have tried to chop my head off and the axe only got half way through my neck. That might be a 10." In a strained, horsy voice I uttered, "eight and a half", closed my eyes and began to reflect on how I came to be there.

Back in 1980, I was 27 and not yet married. For the past 3 years, I had been working as a research scientist for a large consumer goods manufacturing company in the Midwest. I lived alone in an upscale two-bedroom apartment within walking distance from the research center. My life as a single person was pretty much normal. I ate most of my meals out often with my next-door neighbor who was 40 years my senior. I engaged in frequent exercise by jogging several days a week and figure skating at least twice a week. An active nightlife was important to me. I went out most nights often with another single friend from work and had several beers. I also enjoyed scotch. Johnny Walker black label was my favorite. My career was very important to me. Often working late and on the weekends, I wrote research reports and papers for outside publication. I dated on a regular basis but didn't want to commit to any kind of a serious relationship.

You might say that my life was self-centered. I didn't volunteer my time or talents to any organization or cause. Politics didn't interest me. While I was raised a Catholic and attended parochial school for 12 years, I no longer attended church and it had been years since I went to confession. Nevertheless, I didn't see myself as a particularly bad individual. I always tried to keep my commitments. I tried to be brutally honest. I never cheated on my income tax.

My viewpoint of the opposite sex wasn't particularly healthy. In college, I had had a steady girl friend. She betrayed me and my feelings towards her and women in general changed. I found women desirable; however, forming any kind of deep feelings for any one woman was out of the question. My relationship with women had become one-sided and I guess deep down I felt guilty about it but I really didn't care. I thought I was making significant contributions to society from the research I was doing. And, I was fine at leaving it with that.

One particular evening in late February, I was in my apartment watching a college basketball game as March madness approached. All of the sudden, I felt extreme pain across both shoulders as if someone had just broken both of my collarbones with a sledgehammer. I feel off the couch and blacked out for a minute or two. When I came to, I was totally disoriented. The walls seemed to be moving. I staggered to the phone and attempted to call my parents in Philadelphia but I couldn't hit the right buttons. Another few minutes passed. The pain in my shoulders began to dissipate and my head clear. I dialed my parents and my mother answered. I told her I just had some sort of attack and that I couldn't think clearly. For some strange reason, I was overwhelmed with the idea of death. I didn't think that the attack was potentially fatal or anything but I was simply fixed on the notion that I might one day die and the thought was totally unacceptable to me. The feeling was one of genuine fear and I began to tell my mother about it. Of course she responded that we were all going to die one day and that it's simply a part of living that we should choose not to dwell on. After speaking with her for a few minutes my head began to clear but the thought of death consumed me. She suggested that I seek medical treatment but I told her I was going to be all right mainly because I didn't want to alarm her. I ended the call by telling her that I was tired and I needed to go to bed. She stated that I was probably working too hard and just needed some quality rest.

After I hung up, I was tired. I was so tired that I could barely keep my eyes open. I went to bed about 10 p.m. The next morning I got up on time to get ready for work. I felt refreshed but immediately began to think about death. By the time I stepped out of the shower, I was exhausted. I went to work and found it nearly impossible to focus on my lab experiments. I was afraid that a car might hit me on my way home from work and die. Or, maybe I would come down with cancer and die. I wasn't depressed. I could still feel happiness. In fact, I loved the world and everything in it. That was the problem. I didn't want to leave the world. I had become paranoid and completely obsessed with the fear of dying. Dying seemed final. What if there wasn't anything on the other side? Things were wonderful in life. What if everyone was wrong about the existence of life after death? No way did I want to find out. No way!

I got a couple of the lab technicians aside and immediately began talking to them about the existence of God and life after death. I wanted to hear some kind of affirmation. I had to be careful about bringing the topic up. I felt I had lost my mind but I didn't want anyone else to know it particularly anyone at work. While discussing the topic, I found that it relieved my anxiety. I wanted to discuss the topic more but didn't want people to think I had gone off the deep end either so I stopped for the day. For the remainder of the day, I couldn't get my mind off death nor could I shed the overwhelming fear within me.

When I returned home, I was exhausted. I made supper and went to bed about 6:30 p.m. I slept straight through until my alarm woke me at 6:30 a.m. Again, for a brief period, I felt refreshed but by the time I had finished showering, I was exhausted.

For several days, the pattern continued. The fear did not subside. My mind was in constant turmoil. I withdrew from my friends and nightlife in order to sleep. I engaged anyone I could into discussions about the life hereafter. I did find moments of peace. We had just gotten a new capillary gas chromatographic system with a computer. I found that programming the computer to collect and analyze the instrument's data demanded my full attention and enabled me not to think about death. Also, just prior to my episode, I met a new woman who was very attractive. Her name was Marilyn. While I didn't know it at the time, she would become my wife 4 years later. Going out with her was the only exception to getting sleep but I nearly drove her away with my constant ranting of death and the existence of life after death.

In addition to being paranoid and hypersomnolent, I had an insatiable craving for asparagus. I had always detested asparagus. I tasted it as a child but hadn't eaten it since. I could however recall its taste but somehow I was now associating that taste memory as desirable. I also discovered that I had no tolerance for alcohol but not in terms of intoxication. One night after having a few beers, I found myself sitting on the toilet half the night with burning urination. After this happened to me two or three times, I decided that drinking wasn't worth it and basically stopped drinking altogether. So, life pretty much consisted of just three things at that point, living in constant fear, trying to stay awake, and eating asparagus on a regular basis. I don't know how I managed to go to work and remain productive. I found it so hard to concentrate on the mentally demanding scientific work I was engaged with.

I couldn't go on living that way. I considered seeking medical assistance but I was concerned that I would be labeled as mentally ill, which carried a huge stigma with it. I was also concerned that because my employer covered my medical expenses, knowledge that I was mentally disturbed would somehow find its way back to the workplace and my career would come to an end. After all, everyone at work knew Brian had a mental breakdown years ago and hadn't been the same since. I reasoned that I must fear death because I wasn't prepared to die. I needed to do two fundamental things, strengthen my belief in an eternal God and place myself in good standing with him.

After 3 weeks, I decided that I needed to mend things with my faith. On a Wednesday evening after work, I drove to the nearest Catholic Church, went to the rectory and met Father David Garrick. He answered the door and invited me in. I immediately broke down into tears and asked if I could go to confession. After confession, I confided in him what had happened to me. He suggested I see a doctor but I told him my concerns and begged him to help me. He agreed and told me that he would see me every Wednesday evening for as long as I needed. He ended our first encounter by placing his hands on my head and saying a special prayer for healing.

Father David would go on to cure me. He was a most unusual man. Before going into the seminary, he earned a Ph.D. in physics and lived life in the fast lane. One day, he received a call informing him that an intruder had murdered his younger brother in his sleep in his apartment in Houston, Texas. The tragic event was the turning point in his life that led him to the priesthood. He would go on to reveal to me later that he had agreed to help me because I resembled his murdered brother.

My treatment with Fr. David started immediately upon my second visit to him. He focused on scripture to prove the existence of God to me. His knowledge of scripture was truly amazing particularly in his ability to connect the old and new testaments. He showed me evidence of God's presence throughout history in a very convincing manner. Equally important, he emphasized with me the power of free will. He taught me that all I had to do was to choose not to think about death. He started by saying, "Close your eyes and choose not to think about death even if it is only for a few seconds. Repeat this exercise often and try not to think about death a little longer every time you do it." He was right. I could indeed choose not to think about it. Over the coming weeks, I found that I was able to avoid thinking about death for significant amounts of time. He invited me to his Bible study groups to further strengthen my knowledge of scripture. Thus, if I wasn't sleeping, I was either at Fr. David's study group or out with Marilyn.

As the months went by, my fatigue began to lift and my ability to concentrate on other things slowly returned. I continued my free will exercises and programming the computer at work. As I got better, Fr. David suggested that I do some volunteer work. He said, "get your mind off your own troubles by focusing on the problems of others." I agreed and signed up to be a big brother. As I got to better know Fr. David, he confided in me that he was somewhat jealous of me. He talked how God was very much like a true father. I was off doing sinful things and God, like a father, reprimanded me. He explained that God wasn't trying to punish me but that he was just trying to get my attention much like he did to St. Paul (Saul) because he loved me. Father David said that he wished God loved him as much.

One evening, he took the Bible and read a portion of Psalm 18 to me.

The breakers of death surged round about me, the destroying floods overwhelmed me; the cords of the nether world enmeshed me, the snares of death overtook me. In my distress I called upon the Lord and cried out to my God; from his temple he heard my voice, and my cry to him reached his ears. The earth swayed and quaked; the foundations of the mountains trembled and shook when his wrath flared up. Smoke rose from his nostrils, and a devouring fire from his mouth that kindled coals into flame. And he inclined the heavens and came down, with dark clouds under his feet. He mounted a cherub and flew, borne on the wings of the wind. And he made darkness the cloak about him; dark, misty rain-clouds his wrap. From the brightness of his presence coals were kindled into flame. And the Lord thundered from heaven, the Most High gave forth his voice; He sent forth his arrows to put them to flight, with frequent lightnings he routed them. Then the bed of the sea appeared, and the foundations of the world were laid bare, at the rebuke of the Lord, at the blast of the wind of his wrath. He reached out from on high and grasped me; he drew me out of the deep waters. He rescued me from my mighty enemy and from my foes, who were too powerful for me. They attacked me in the day of my calamity, but the Lord came to my support. He set me free in the open, and rescued me, because he loves me.

He read it to me because he sensed my extreme level of despair and wanted to assure me that God would come to my rescue if for no other reason than he loved me. This psalm would come to have special meaning to me over the years particularly when Chiari would come knocking again later in life.

Over a nine-month period, the fear, paranoia, and hypersomnolence gradually lifted. However, my preference for asparagus remained which I consider a benefit today. Not only did my concentration return but also my work improved significantly. I had the opportunity to speak at a world conference in Acapulco on Soy processing and utilization and followed it up with one of the most comprehensive publications in the scientific literature on the analysis of Soybean oil. Also, while at the conference during my free time, I tried paragliding from the back of a motorboat. I recall sailing past the tops of the high-rise hotels along the beach and thinking that my fear of dying was truly conquered.

So what really happened to me on that February evening? I was stricken with intense pain followed by a brief blackout then left with longer-term problems of paranoia and hypersomnolence. Further, my taste radically changed. It doesn't fit the profile of clinical depression. While I did not seek medical attention, I did call my brother who is a physician. He agreed that it did not sound like depression primarily because I was readily able to experience happiness. He suggested that something organic might have happened like a mini-stroke. While there is no way of knowing for sure, I can't help but wonder if it wasn't the first signs of CMI. In the preface, the World Arnold Chiari Malformation Association or WACMA is mentioned. WACMA is not an organization. It is basically a web site with information on CMI. Associated with it is a support group on Yahoo egroups. Over the past five years, I have read thousands of messages posted on Yahoo by Chiarians and have exchanged personal email messages with hundreds of Chiarians. People seeking to confirm the various strange problems and feelings they are experiencing often posted on the topic of symptoms. While it is not possible to decipher what symptoms are due to CMI or syringomyelia (SM) and which are due to medications and/or concomitant illnesses on-line, one does, nevertheless, over a period of several years come to believe that almost anything can happen with CMI. I have also communicated with other adult Chiarians in their forties and fifties who claimed to have had neurological problems earlier in the life time that vanished as mysteriously as they appeared.

Speculating further on this, it is conceivable that some cerebral spinal fluid (CSF) blockage initially occurred causing compression and damage to some region of the brain, which over time may have healed itself. By that, I mean, since damaged neurological tissue doesn't regenerate, that the brain found another pathway to process the neurological tasks. Another possibility is that the blocked CSF forced its way into the spinal cord to form an asymptomatic syrinx that relieved the pressure in the brain and stabilized the individual for several years until the condition worsened again in later life.

At any rate, I got better, physically, mentally and spiritually. I made important and necessary changes in my life. I was attending church on a weekly basis. I was no longer interested in superficial relationships. I stopped drinking. I was volunteering my time and talents to a worthwhile cause. I resumed my friendships. Work and my career were back on track. I traveled through illness and became a better human being as a result.

Still, I was left shaken by the experience mainly because I didn't know its cause and I wondered if it might some day return. While I had emerged a better human being, I also felt robbed of nine months of my life. I confided in my closer friends about what had happened to me and told them that I would rather lose my legs than go through the mental torment and fatigue again. And, I meant it.

Chapter 2
An Extended Hiatus

As a result of all the sleeping, my regular exercising was discontinued for an extended period of time and I gained weight, about 40 pounds. I carried the weight for a couple of years. Being 6'5", I could get away with it so to speak but it bothered me. One day, we got into a discussion of weight and dieting at work. It ended up in a bet with an associate of mine as to who could lose their excess weight the fastest. I went on an extreme diet of only 500 calories a day. My daily food intake consisted of black coffee in the morning, a bowl of vegetable soup at lunch, and a turkey sandwich and apple for dinner. No mayo of course. On top of that, I jogged 3 miles a day. The ketone level in my urine was off the scale and I was cold all the time. It was summer and I was going to bed with blankets. But, the pounds flew off and within 6 weeks I lost over 45 pounds. This diet was of course too extreme and I was lucky that didn't do damage to my health but bets have forced people to do stranger things I suppose. I recall well the first morning that I came off of the diet. I ate a glazed donut and experienced a true sugar high.

After losing the weight, I resumed my running with a new goal of completing a marathon. I trained for several months and at 29, I completed my first marathon in three hours and thirty-three minutes, which equated to a pace of about eight minutes per mile. I mention it because when I went home and took a shower, my arms tingled so intensely that I thought something was happening to my heart. The tingling stopped after a few minutes and I chalked it up to overexerting myself. I ran another marathon the following year and did not experience any tingling afterwards. Nevertheless, it could have been a warning sign of CMI as tingling of the extremities is a common symptom.

In retrospect, I have wondered about the risk of running for the non decompressed Chiarian. To understand why, one must first have a basic knowledge of cerebral spinal fluid flow. Earlier, it was stated that CSF drains from the skull into the spinal canal but it is important to understand the nature of this flow. It is not a slow continuous type of flow as intuition might suggest. Rather, there is a pulsating pattern to the flow. When the heart contracts it sends blood up to the head and temporarily swells all the many fine arteries and capillaries in the brain. This increases the volume of brain tissue but the volume of the skull is fixed. To compensate for the fixed volume of the skull, the clear fluid in the brain, the CSF, leaves the skull and enters the spinal canal. When the heart expands, the flow of blood reverses, the volume of brain tissue decreases and CSF is pulled up into the skull from the spinal canal. Thus, CSF flow is actually pulsatile in nature. It pulses back and forth across the foramen magnum (the large opening at the bottom of the skull) with each heartbeat.

The amount of CSF volume that is forced down into the spinal canal when the heart contracts is slightly greater than the volume that is pulled back up when the heart expands. Thus, the net flow of CSF is in favor of leaving the skull and draining down the spinal canal. One can essentially think of the spinal canal as a reservoir for the skull that keeps the delicate pressure on the brain more or less constant. With the cerebellar tonsils positioned in the foramen magnum, they too pulse with every heartbeat and experience a net drag in the direction of the spinal canal. This has been directly observed with both Cine MRI and ultrasound. This increase in the pulsation rate of the tonsils over an extended period of time cannot be good. Jogging and increasing heart rate after decompression, however, should no longer be a problem, as CSF flow is no longer blocked across the foramen magnum (see appendix 1 for further understanding).

Life in my thirties unfolded like the American dream. I married Marilyn and had a son whom we named Dominic. I was transferred to the Company's pharmaceutical division in New York and my career took off. I was very active throughout my thirties and didn't experience a single neurological symptom. While in upstate New York with a new assignment and a young family, I didn't have time to jog. Also, the roads through the many dairy farms in the area were filled with many overly protective canines, which made jogging in the neighborhood, somewhat a hazard. However, I did engage in heavy work. Our property was in the midst of a heavily wooded area. I felled trees and bucked and split large quantities of firewood all by hand for our air-tight wood burning stove. At our elevation of 2000 feet above sea level, we typically accumulated a hundred or more inches of snow each winter. We had a 150-foot driveway, which I would shovel by hand every time it snowed. This heavy work also increased my heart rate on a regular basis but I did not experience any problems, which serves to fortify my earlier theory that my CSF had found a way to drain perhaps via an asymptomatic syrinx.

In 1992, at the age of 40, I was transferred back to the mid-West in the same capacity. With a familiar job, more independent children and more user-friendly roads, I resumed my jogging. I had gained back about 20 pounds and was interested in trimming down a bit. At first, things were fine. Not having run for awhile and being somewhat older, I experienced some minor tendonitis, which was manageable with over-the-counter non-steroidal anti-inflammatory drugs like Motrin or Advil and backing off a little on the intensity of my running. I would usually run six days a week and take Saturdays off. On a typical Saturday, I would take a short nap in mid-afternoon. After about a year, I noticed when trying to take my Saturday nap, an uncomfortable sensation in my legs that began to inhibit my ability to fall asleep. The closest description of the sensation is that it felt like thousands of insects crawling on my legs but since I never had thousands of insects crawl on my legs, I'm not really sure about the accuracy of the analogy. I didn't know what to make of it. I didn't experience it at night so I chalked it up to getting older, ignored it, and stopped taking naps all together. Right or wrong, I had heard that you required less sleep as you aged and I thought it was beginning to happen starting with the need not to nap. I did not initially connect it as any kind of neurological problem or symptom.

My wife also noticed that my voice would get hoarse after doing yard work starting at about the same time. She would say, "Are you catching a cold?" to which I would reply, "No, it's probably just a little irritation from the dust I kicked up". This would turn out to be a very interesting neurological phenomenon as time went on. I was also noticing while watching television during the evening with my family that I had a need to take deep breaths on a frequent basis as if I was running out of air. I knew that it was perfectly normal to take an occasion deep breath. In fact, I remembered hearing a lecture once by a biology professor doing research on breathing in lizards that we all needed to take deep breaths on occasion to align the microscopic cilia in our lungs. But, I found that I was taking deep breaths every couple of minutes and I began to suspect my eye drops for my glaucoma. I knew I was taking a selective alpha agonist and that such medications could have cardiopulmonary side effects. I checked my PDR (Physicians Desk Reference), which indicated shortness of breath as a possible side effect. The following day, I called my ophthalmologist and asked if the shortness of breath I was experiencing could be due to my medication. He indicated that it could but that he rarely saw it and instructed me in the proper way to take eye drops so as to minimize the amount of drug that flows down the tear ducts to increase systemic concentration in the body. I began to take my drops as directed – placing my head down after administering the drops, lightly pinching the corners of my eyes across the bridge of my nose, and wiping away the excess fluid after 1 to 2 minutes. I still found myself taking frequent deep breaths even after trying this for several days. It continued for a few weeks then went away as mysteriously as it had appeared.

On a very personal note and for the sake of being complete, another unusual change was taking place that had to do with my sexual function. Immediately after climaxing, I would experience intense burning in my penis identical to that which made me decide to stop drinking several years ago. As a result, acts of intimacy with my wife became infrequent. Like the strange sensations in my legs, I interpreted the problem to be age related. I also began to lose interest in sex and found attractive women in general to be less distracting. These changes bothered me to the point where I scheduled a physical with my primary care physician. I didn't talk to my doctor about the burning sensation during sex. A normal prostate exam and PSA level was good enough for me. I did however discuss my lack of interest in sex with him. He asked me if it bothered me. I told him no. In fact, I went on to tell him that I found it to be quite liberating. He indicated that he had never heard that response and asked me to elaborate. I told him that all my life, I found women very attractive and it had often got in the way of me concentrating on my studies or work at hand. I explained that it was no longer a distraction and I was better able to focus in general. I was in essence free to focus. Also, I never felt at peace with myself. I felt that my fantasies were a superficial reflection of my attitude towards women. I felt that I should respect women more for what they were as people as opposed to objects of desire. Thus, for the first time in my life, I found that I could interact with women in a more genuine manner and I liked it. It made me feel free. Needless to say, my doctor didn't know how to respond except to tell me that he had seen other older men lose interest and if it didn't bother them, it wasn't really a problem.

As time went on, I decided that I wanted to run another marathon. Training up from where I was would be relatively easy, or so I thought. I was already running about 6 or 7 miles a day. Going on my experience from my late twenties and early thirties, I figured it would only take a couple of months to get to 10 miles a day. I began to train but found that I wasn't improving to any real extent. Again, I had heard that running and sports in general became more difficult after the age of forty. So, I dialed back my expectations. I also noticed that my stomach was often a bit queasy after jogging. Again, I tossed it aside as age related and began taking Tums on an intermittent basis.

The things that were happening to me were early warning signs but I was confusing them with natural aging and really didn't give them a second thought. I never ran the marathon and kept making excuses to myself like I didn't have the time to take the extra training that I needed.

Chapter 3
The Hawaiian Surprise

It was late January 1998, and I was on an international flight to England on business. Like all my past transatlantic trips, I knew I wouldn't get any sleep on the plane. I was never able to sleep on planes. There are just too many distractions and jet engines are just too noisy. I never understood how some people managed to do it. Of course, those who fall asleep easily on planes always seem to be the ones that snore the loudest as well and contribute to the noise preventing me from sleeping. Anyway, I enjoyed flying business class and saw it as an opportunity to see some movies. My travel schedule had been busy that month with prior trips to Canada and Arizona. My usual practice was to arrive a day early, check into the Runnymeade Hotel around 11 a.m., take a nap for a couple of hours and then go out for a run along the banks of the Thames. The Runnymeade was booked so my secretary had arranged for me to stay in neighboring Windsor.

I arrived at the hotel in Windsor just before noon. My room wasn't ready so I sat in the lobby for about an hour. When I got to my room and went to take a nap there was a lot of racket coming from the window. I looked outside to see that the hotel's patio was under construction. Since there was no way I could fall asleep with all the noise, I decided to skip the nap altogether and walk around Windsor to locate a restaurant for dinner that evening. While on my walk, I stopped for coffee to help with the jet lag. I found an interesting

Italian restaurant right across the street from the castle and returned to my room to change into my running gear. After my workout, I showered and went to the restaurant for dinner. I had a dandelion salad and some sort of seafood pasta dish. While eating the pasta dish, I came across a mussel that didn't taste quite right but ended up swallowing it before I realized it. I made a mental note of it in case I got sick later.

I returned to my room after dinner. The work outside my window was done for the day and it was quiet at last. I decided to stay up however and watch television in order to force my biological clock into synchronization with the five-hour time difference. I went to bed about 10 p.m. By then I had been awake for 34 hours and was exhausted but found it impossible to fall asleep. I finally fell asleep about 4 in the morning but had to get up at 6 a.m., so I only got about two hours of sleep. While I was tired, I wasn't concerned. I reasoned that the time difference was the cause and I would eventually accommodate to the change during the week as I always managed to do on previous trips.

I arrived at the office about 7:30 a.m. and headed straight to the coffee machine. I drank several coffees during the day to stay as alert as possible while I met with my subordinates. That evening, I went to dinner with one of my managers. We had a couple of cocktails with some quail eggs before the meal. I got back late but again I found that I could not sleep. I wasn't worrying about anything. There were no pressing business issues. I was on a routine trip to meet with part of my global organization. I got up and dressed and went for a leisurely walk along the Thames late at night. I never did sleep that night and once again I went to the office the following day and consumed copious quantities of coffee.

For the 5 business days that I was in England on business, I figured I slept only about 5 hours. For sure, things would return to normal once I got home and got back to my routine.

While on the way home, I thought about making vacation plans. I had been working hard and traveling a significant amount of the time and wanted to spend some quality time with my family. I decided that once I got home, I would announce that we would be going to Hawaii during my son's school break in April. I had accumulated enough frequent flyer miles to take a dream vacation. When I returned home, I initially found that I was able to sleep. After catching up on my sleep, I sprung the news about the vacation plans. Needless to say, my wife and son were in favor. The very next evening, we had plans to attend dinner at a friend's home. That's when I first began to notice pain in my throat upon swallowing.

The following week, my insomnia returned and I began taking Tylenol PM to promote sleep. The pain during swallowing became much sharper. The odd thing about it was that it only happened during dry swallows. There was no pain at all when swallowing food or drink. I thought that I must have picked up some sort of strange respiratory infection probably on the plane during my trip. Being the type that doesn't jump to the doctor for every little thing, I decided to wait and see if the sore throat would go away without medication. I continued to run that week, which usually promoted sound sleep but was still having trouble falling asleep and would take Tylenol PM as needed.

The next week was a particularly busy one at work. My department was having an off-site global technology strategy setting meeting. All of the managers in my department along with selected technologists were meeting to evaluate and select a new technology platform that we would implement for collecting data from clinical studies around the globe. During the first day of the meeting, I was finding it difficult to concentrate on the presentations and discussions. I kept swallowing like a criminal on the hot seat during questioning. Each time I dry swallowed, it felt like someone was jabbing a knife in the left side of my throat. It was extremely painful. It was like no other sore throat I had before and I was totally mystified that it wasn't sore at all when swallowing food. During the meeting, I decided that I would call my physician if I still had the problem the next day.

When I woke up the next morning, the first thing I did was swallow. The pain was still there. That day I broke away from the meeting for a few minutes to call my doctor's office. My doctor wasn't in but his associate was available to see me after work so I booked an appointment for 5 p.m. At the end of the meeting, I went to the doctor. I explained that I had recently taken an international trip and picked up some kind of sore throat, that I had it for a week and a half and it wasn't getting better. He checked me out and said that he saw no inflammation in my throat. He prescribed Ceftin, an anti-infective, and instructed me to make an appointment with an ENT if my throat didn't feel better in 5 days. The unusual soreness plus the fact that he didn't see any inflammation raised the level of my concern. I stopped by the pharmacy on my way home and got the prescription filled.

My throat did not improve after taking Ceftin for five days and my left ear was beginning to hurt. I called my doctor's office again and asked to talk with my physician. He had returned and the receptionist indicated that he would call me as soon as he got a chance. When he called back, I told him what was going on and asked if he would prescribe a different medication just in case my infection was due to an organism resistant to Ceftin. He hesitated but gave in and prescribed Biaxin and told me to see the ENT if it wasn't feeling better after a few days. I knew there was something going on other than an infection but I guess I didn't want to find out. I wanted it to be an infection like strep-throat and I was hoping Biaxin would knock it out. My secretary was telling me that her daughter had a sore throat as well that they were having trouble knocking out. Likewise, one of my managers, Don, complained to me that he too was having throat pain when he swallowed. When he told me this, I questioned him as to the nature of it. He told me that when he swallowed it felt like something was sticking in his throat. He went on to tell me that he had seen his doctor who had ordered some tests. Don didn't seem to be the least concerned. After hearing these two stories,

I rationalized that some bug was going around and I had it as well. A few days passed and there was no improvement. In fact, my left ear was hurting even more. I called the ENT and made an appointment for later in the week. While waiting to see the ENT, I tried to keep my mind off the pain but it was difficult. I found myself swallowing more and more to the point where I lost perspective of what a normal swallowing rate was. Analgesics provided no relief – another sign I didn't like.

When I got in to see the ENT, I explained what was going on. He seemed very attentive and genuine. Before he examined me, he explained that throat pain is often referred to the ear. After he completely examined me, he told me that he really didn't see anything that was obviously wrong. He did however notice the pooling of a small amount of saliva on the left side of my trachea due to muscle spasm. He explained that no saliva should be left in the trachea after swallowing and went on to say that he could only think of two things that could cause it. "The first is a tumor under the larynx where I can't see it. However, that isn't likely because you aren't a smoker. The second possible cause is an upper esophageal burn from night time silent GERD (gastro-esophageal reflux disease)." He explained that some individuals without GI complaints some times have GERD and don't realize it. So, he ordered a CAT scan of the neck to test for a tumor and a barium swallow to test for GERD. His nurse called the nearby hospital and scheduled the tests. I would have to wait another week.

By now it was the end of March and our trip to Hawaii was fast approaching. I drove home from the ENT visit thinking that I'm going to find out that I have throat cancer and it's going to ruin our vacation plans not to mention my life. When I got home, I told my wife the outcome of my visit. "I'm really concerned about the possibility of cancer because the nature of the pain is so bizarre." I also told her how much I was looking forward to the vacation and that I didn't want to disappoint her and my son. She tried to calm me down and told me not to worry if we had to cancel the vacation. Later that evening, I called my brother, a pulmonologist. He said it was probably nothing to be concerned with and echoed what the ENT had said about the association between throat cancer and smoking. I felt better after speaking with my brother, popped a Tylenol PM and went to bed.

I found the following week waiting for the tests unnerving. My throat and ear continued to hurt. It's nearly impossible not to think about pain that is right in your face. I was beginning to show signs at work that something was brothering me. Another one of my subordinates, Burt, who I was particularly fond of, could read me better than most people. He asked me "Is something wrong, Ray?" I responded, "I went to an ENT who suspects throat cancer as one possibility and GERD as another. I'm very concerned because of the nature of the pain I'm experiencing and the fact that I had no history of any GI problems." Burt was 10 years my senior and a very level headed manager that I had come to greatly respect. He offered to take work off my shoulders if I felt it would help. I would come to take advantage of Burt's offer as time went on.

Getting quality sleep was difficult and I wondered if the lack of sleep was somehow contributing to the problem. The day for the tests finally arrived. I went to the hospital in the morning and didn't have to wait. The CAT scan was performed first. I was to be scanned with and without contrast. An IV was inserted into my arm for administering the contrast agent. I recall feeling warm immediately following its injection - an expected reaction. When the test was over I was given a copy of the scans to take back to the ENT. The barium swallow followed and the radiologist performing the test told me that things appeared normal. That was good news of course but that left the possibility of cancer. I was to return the following day to the ENT with the scans. As it was mid morning, I returned to work to finished out the day. I parked my car and upon walking to the entrance went past the designated smoking area where a couple of younger people in my department were smoking. I greeted them cordially but deep inside I wanted to reprimand them for smoking and remind them of the risk of cancer. It didn't seem fair that I, who never smoked, should be faced with the risk of cancer and they who smoked routinely, were apparently healthy and didn't have a concern in the world. That evening, I looked at the scans but didn't have a clue as to what a tumor might look like.

The following afternoon I returned to the ENT with the scans. My wife accompanied me for moral support. I was quite nervous about the possibility of having throat cancer. I read up on it and learned enough to scare myself. He placed the scans on his viewing box and looked at each one very carefully. He then pointed to a particular scan and said, "See that there? That's were I would expect to see a tumor if there was one but there isn't. You're fine". I thought that I would have been relieved to hear negative results but some strange reason my reaction was one of ambivalence. I was glad that no tumor was found but on the other hand, my throat and ear were still extremely painful. I asked why I was still in pain. He stated, "The only thing that could be causing the muscle spasm in your throat at this point is stress and you should go on vacation and no longer worry about it."

Friday, April 9th arrived and we departed for Maui. As frequent flyer award trips often are, our flight was a long indirect one with a stop in Dallas and Honolulu. By this time, I not only had soreness upon swallowing and ear pain but excessive post-nasal drip as well. I had no idea what was causing the post-nasal drip because I had no allergies and did not feel that I was getting a cold. All the way over, I kept swallowing and swallowing. And with each swallow, I could feel that sharp knife pierce my throat. When we arrived at the airport in Honolulu, we were given the traditional Hawaiian greeting and preceded to the gate for our final leg of the journey to Maui only to learn that the flight was delayed 5 hours.

We decided to go to the cafeteria in the airport and get something to eat. Eating was a good thing as I felt no pain when swallowing food. Eating helped take my mind off the problem. During the days leading up to the trip,

I thought about stress. Perhaps I was stressed out and didn't know it. I said to myself, "I did assume significantly more responsibility over the last year and now I have a total of over 110 people under me with 16 direct reports. I don't want the assignment. I took it only because the vice president pleaded with me." Nevertheless, there wasn't anything about it that was particularly problematic. It carried a large administrative load but I had a great assistant who kept me on track.

I thought about my personal life. I was happy with my marriage. My son was doing well in school. My stepdaughter who I raised from the age of 12 was happily married. I had always had a terrific relationship with my parents. The only thing I began to question was my career, not in terms of the level of responsibility but in terms of the nature of the work. I was trained as a chemist but somehow drifted away into the world of clinical research. Nevertheless, I found the work interesting. Net, I couldn't identify anything significant that I would consider as stress causing.

We arrived in Maui late, about 10 p.m. We had been traveling for 19 hours. By the time we rented a car, drove to the hotel and settled into our room, it was after midnight. I was continuing to have difficulty sleeping. I was taking either Tylenol PM or Benadryl every night. After unpacking and changing for bed, I asked my wife for medication and went to bed. I was exhausted and fell asleep relatively quickly. About 4 in the morning, I woke up. I was nervous and crying on the edge of the bed. My wife awakened and asked me what was wrong.

I told her that I thought I was losing it again like I did when I was 27. I told her while crying that I couldn't decide what we should do or see that day. The thought of having to make what would normally be considered a simple decision seemed insurmountable. She calmed me down by telling me that we didn't have to do anything; that we could just hang around the hotel and relax. It seemed to work and I stopped crying. It was Saturday, April 10th, the day before Easter. As the day went on, I began to feel better emotionally but the pain in my throat and ear persisted. After lunch, I suggested we get in the car and drive around the island just to get acquainted with things. The drive afforded many opportunities to view beautiful beaches and coves. We had brought with us on the trip gear for snorkeling and we were looking for just the right place to go the following day. Again, that evening, I took a double dose of Benadryl and went to bed about 11 p.m.

On Easter Sunday morning at 4 a.m., I woke and immediately began crying. Again, my wife heard me and woke up. I was holding my head and told her that something was tearing inside my brain. I didn't have a headache. The part of my mind responsible for rationale thought was being stripped away against my will. My mind was in complete turmoil and all I could think of was going to the balcony and jumping off. I told my wife that I was having a breakdown and thoughts of suicide and asked her to take me to a doctor. She called the front desk to see where the nearest doctor was located. We would have to go all the way to the other side of the island to the hospital. She got directions and woke up our son. By the time we were ready to go, I had urinated three times. I was urinating just about every 20 minutes. After we were all ready, we went out to the car and my wife set out for the hospital.

While driving, my wife tried to talk to me. She asked me what was wrong. I told her that the pain was driving me crazy and that I was convinced that the doctors had missed something seriously wrong. I told her that there was just too much pressure on me. I didn't know what to do with the family and that we should probably catch the next flight home after I got some sedatives to relieve my anxiety. All of my life, I had tried to avoid taking medications unless essential as in the case of an infection. Once, while in college, I got stressed out and went to the campus physician. She prescribed Valium for me. I took one or two doses and didn't like the way it made me feel. It dulled my mind but didn't address the root cause of my anxiety so I stopped taking the medication and just decided to stop letting things bother me. Sometimes making firm decisions is the best medicine. This time was different. I was incapacitated. I couldn't make the simplest decision and suicide seemed so desirable. Before we left the hotel room, I looked off the balcony and could only imagine peace if I were to jump off. I tried to summon my will power as Father David had taught me but it was nowhere to be found. As much as I hated medication, I knew it was the only course of action.

It took about 45 minutes to reach the hospital. Upon arriving, I immediately headed to the men's room to urinate and then went to the front desk in tears with

Marilyn. After the usual filling out of forms and providing proof of insurance, I was taken immediately to an exam room as no other patients were in the emergency room. A nurse asked me what was wrong. I told her I had a nervous break down and that my throat and ear were in a great deal of pain. She could see that I was unable to stop sobbing and was very compassionate. When the attending physician arrived, I told him about my swallowing problem and sore ear and insisted that something was very wrong in my head. He examined me and noticed post-nasal drip and inflammation in the back of my throat. He diagnosed me as clinically depressed with an allergy and prescribed Ativan, a sedative, Paxil, an antidepressant, and Flonase, a topical steroid for allergies. He recommended that we not return home but rather that I stay in Maui and try to rest as the islands where a very special place of healing. He said that the Ativan would help me cope in the near term until the Paxil kicked in and there was no reason to cut our vacation short. It sounded logical to me and I really wanted my family to have a vacation so we decided to stay.

Before leaving the hospital, Marilyn got directions to the only pharmacy open on the island on Easter Sunday morning to get the prescriptions filled. While in route to the pharmacy, I thought how I had disappointed my family. I had completely ruined our big vacation. Marilyn got the prescriptions filled and brought them to me in the car. I immediately took some Ativan. In just a few minutes, I was feeling considerably better. I stopped crying and the urge to urinate subsided. At that time, I knew little about antidepressants. I had the impression that they were serious medications with significant side effects. I told Marilyn that I wasn't going to take the Paxil until I saw a doctor at home. While the Ativan was effective at relieving my anxiety, I remained seriously depressed with thoughts of suicide. When we returned to the hotel, Dominic turned on the television. I stared at the television unable to follow the plot of the show. After an hour or so, Marilyn suggested that I might feel better if I went jogging. I agreed. Running always helped to clear my head. I changed into my jogging attire and went for a run. I immediately found it difficult to jog. My legs felt heavy and I got winded after a short distance. I should have been able to run 6 or 7 miles with no problems but I returned to the hotel after running only one and a half miles. I wasn't sure if it was the Ativan or my illness. While on my run, I came across a small public beach. Upon returning to the room, I suggested that we go to the beach so that Dominic and I could go snorkeling. Marilyn didn't swim. We went to the beach and Dominic and I went snorkeling for the first time. Not only did we experience a variety of beautifully colored fish but we could also hear the clicking of whales that were migrating past the island. While unable to feel any kind of joy or happiness at all, I was pleased that my son was having fun and that my family would be able to enjoy their vacation after all.

The Ativan did in fact stabilize me and allow me to sleep. It enabled me to make decisions on where to go and what to see, with the input from my wife and son, of course. That week we saw all that the island had to offer but I personally didn't take joy in any of it.

Concurrent with my acute depressive episode, the pain in my throat and ear spread to the left side of my face, jaw, and neck, and the little finger on my left hand began to tingle. The pain that extended from my throat to my left ear now seemed to travel along an axis deep inside my pharynx. I decided that when I got home, I would first see the ENT again because I was certain there was still something physically wrong in my neck or head and then go for psychiatric care.

Chapter 4
Nightmares

When we got home, I returned to the ENT as soon as possible. I told him how the pain had spread. He was very patient with me and once again examined me only to confirm that nothing was wrong. He even gave me a hearing test, which showed that my hearing was normal despite the pain in my left ear. When I told him about the tingling in my finger, he exclaimed that that wasn't in his ballpark and that I should see an internist. Since my primary care physician was also an internist, I made an appointment with him. I also called my health care plan's number for mental illness assistance and made an appointment with a counselor. All the while, I was in great mental distress. Sleep was getting difficult despite the fact that I was taking a sedative. I could not concentrate, not even enough to read a newspaper. Going to work took all the will power that I could muster. While at work, I couldn't follow simple logical arguments made by others in meetings. At home, I was able to let down my guard. I cried a great deal. I saw no value in anything I was doing. On an intellectual plane, I did appreciate my responsibility for Dominic's well-being and development. I was concerned that he had been with us when I broke down. I took him to the park for a walk along the nature trail. The first thing I did was reassure him that I would be all right and then went on to explain the importance of recognizing the totality of his being. That he was a being composed of a body, a mind, and a spirit and that all three required nourishment. If one were neglected, the other two would also suffer the consequences. I told him that I had been neglecting my spirit and now my mind and body were paying the price but I would turn my attention to my spirit and get well. He grasped what I was saying and I changed the subject to suit his interests.

The next day, I went to the mental health councilor. Her name was Ramona. Ramona was middle-aged and very empathic. As I began to tell her my story, I began to weep. As hard as I tried, I couldn't hold back the tears. Ramona could tell that I was clinically depressed and in need of professional help. She recommended that I see a psychiatrist and gave me a list of about 6 names to choose from. Then she suggested that I seek the services of a spiritual advisor. At that point, Ramona revealed that she was a Catholic nun. I responded that I was surprised that a nun would be working for a secular organization. She explained that it wasn't uncommon and consistent with both her educational background and calling. She told me that she had a spiritual advisor and found her very helpful. I took the time to tell her about Father David but he moved out of the area many years ago. I also told her that I was hesitant about approaching my pastor or any of the other priests in residence at my parish. Ramona indicated that she would talk to a priest for me that she knew and would call me back. When I got home, I told my wife about the plan and selected the psychiatrist on the list who practiced nearest to our home. I had no other basis to choose from and didn't want to drive across town. I called the psychiatrist's office and was able to get an appointment the very next day at 5 p.m.

I did not get a good first impression of the psychiatrist at my appointment on the following day. As they say in the business, the transference was bad. He came off as polite but very clinical. He asked me several questions. "Do you have trouble concentrating? Are you thinking of suicide? Do you hear voices?" I answered his questions. "Yes, I'm having trouble concentrating. Yes, I think about suicide a lot. No, I do not hear voices." I then told him what had happened to me with the onset of my physical symptoms and my episode in Hawaii. He acted like he had heard it a thousand times before and said he could help me. I told him that I was convinced that the medical doctors had missed something important. He said, "You can go to the Mayo clinic, let them turn you inside out, and nothing wrong will be found." I thought it was very presumptuous of him. I went on to tell him that the ER doctor in Maui had prescribed Paxil for me but I hadn't taken any yet. He said that he preferred Prozac but Paxil would be fine and instructed me to begin taking it at twice the dose. I told him that I was really having trouble sleeping and that my problem with insomnia had started several weeks before the onset of the depression. He simply responded that the Paxil would take care of that after a few weeks. He finished by telling me that his associate in the office next door was available for psychotherapy sessions, which I should have twice a week and that he would follow my progress and adjust my medication.

I made appointments for psychotherapy with the receptionist and left the office. I didn't want my sessions to interfere with work and they accommodated my needs by making my appointments at 5 p.m. or later.

I began seeing the therapist and taking my Paxil. The Paxil gave me abdominal pain and seemed to only make my insomnia worse. In addition, the pain in my face, jaw and neck began to migrate around. It would focus in on a specific area for about 3 days and then move to another. The pain would even occur in individual teeth. In addition to pain, my cheek began to go numb and the entire left side of my face felt like it was sun burned. Sometimes even the left side of my tongue would go numb.

The therapist seemed nice enough but the sessions didn't seem to be relevant to me. We honed in on the topic of work and how the turns in my career carried me away from my original interests. I'm sure he thought he was on to something. I remained certain that the origin of my problem was physical. When I next saw the psychiatrist, I complained about the GI pain and the worsening insomnia. He increased the dose. At the higher dose, I was sleeping only about every third night and then not more than 4 hours. I looked up Paxil in the PDR and noted that it indeed caused insomnia at about the same incidence that it caused somnolence. I did some searching on the web about Paxil and insomnia and confirmed that, in general, SSRIs (Selective Serotonin Reuptake Inhibitors), like Paxil, Prozac, and Zoloft, often caused insomnia.

One day after two consecutive nights of no sleep, I called the psychiatrist and complained again about the severe insomnia. He called in a prescription for trazadone for me. I immediately looked it up, saw that it was the most sedating of the old tricyclic antidepressants, and was anxious to give it a try. That evening, I took the trazadone. After 45 minutes or so I began to feel very drowsy and I went straight to bed but then I became very congested and couldn't breathe through my nose at all. It was the worse case of nasal congestion I had ever experienced. I came down stairs and looked up trazadone again. Sure enough, nasal congestion was listed as a possible side effect. I tried to sleep that night by breathing through my mouth but my attempt failed. Fortunately, it was a weekend and I didn't have to go to work. By mid morning my congestion cleared up and I promptly flushed the trazadone down the toilet.

Ramona called me back. She said that the priest she had in mind for me as a spiritual advisor had just been transferred out of town. However, she went ahead and talked to another priest by the name of Father Al Bishoff, a Jesuit at a local university, about me and he was willing to see me. I called Father Al and introduced myself. He indicated that he would be willing to see me every Wednesday evening at 7 p.m. until I was feeling better. Father Al was in his late seventies and served as a student advisor on campus. I liked him from the very start. I told him about my past and how Father David had saved me from paranoia. I went on to explain that ever since my awakening 20 years ago; I had trouble with the concept of prayer. I explained that while at the height of my arrogance, God struck me down. I had tried to stand in the center of the universe and learned the hard way that only God could withstand the infinite forces at the center of the universe. I was awestruck by the experience and at God's majesty. I didn't feel worthy to pray to him or ask him for favors or healing again. I had gotten away with too much and didn't deserve any further attention. He healed me through Father David and that was more than I deserved. Father Al said, No, you are completely wrong about this. If you are having trouble with the almighty nature of God the Father, take your mind off him and ask Jesus, our brother, to help you." Father Al guided me back to prayer and I would continue to see him on a weekly basis.

That weekend, I had a regional science fair to judge. I had agreed to do it at the beginning of the year before I became ill. I wasn't looking forward to it but felt it was important to the kids to keep my commitment and it might help keep my mind off my troubles for a few hours. I was concerned however that I might have difficulty following the projects of the older students. The event was being held at a university about 45 miles away and judging was to begin at 9 a.m. I left early and still ended up parking a good distance away from the building. I went to the judges' room and was assigned to the 7th grade level. I was thankful, as I knew I could handle that. After completing my judging, one of the organizers asked me to judge a special award at the senior high school level. I bit my lip and went to the room with two other judges. We had to judge three very good projects. I managed to focus and get through all three with a reasonable appreciation of what they were about. We returned to the judges' room and discussed the projects to select the winner. I wasn't able to contribute much. I made some comment about separating what the student had truly accomplished versus the obvious adult help that some of them had had. After considerable deliberation we selected the winner and I was free to go. In my car on the way home, the sharp pain upon swallowing began to go away. I was amazed but the pain and numbness on the left side of my neck and face remained as well as the deep pain in my left ear. That night while working in my study, the pain in my ear went away very abruptly. Something had clearly changed but not all and the tingling in my little finger crept up my arm to my elbow. Also, my Eustachian tubes began to pop very loudly with each swallow. While I had heard that this was common for some people, it wasn't for me. And, when I swallowed, the left side of my throat felt like it was moving slower than the right.

The next weekend was our annual mulch project. I had had 5 cubic yards of bulk mulch delivered for spreading around the various trees and scrubs on our property. As I began to shovel the mulch into the wheel barrow, I began to sweat intensely which I understood was a common side effect of Paxil but I also discovered that my arms were weak and a peculiar sensation of choking around my neck had developed which I did not anticipate. However, when I looked up Paxil in the PDR to check, weakness and tightness were also listed as possible side effects. Thus, while I didn't yet realize it, I began to get caught up in the very problem that makes diagnosing Chiari so difficult. Many neurological symptoms can be caused by depression or the very drugs to treat depression or any number of neurological diseases including Chiari.

Late May rolled around; I had been taking Paxil for several weeks. There was no improvement in my mood; the pain was ever present and the insomnia nearly unbearable. I decided to talk to Marty at work. Marty was an internist and knew me well as a friend and coworker. He took my complaints seriously and said that he wanted to think about it for a while. The next day, Marty came to my office early in the morning. He told me that he thought he had recalled something from medical school and looked it up at home the night before. He was fairly certain that I had a case of glossopharyngeal neuralgia (meaning pain of the 9th cranial nerve). Marty explained that glossopharyngeal neuralgia was caused by compression of the ninth cranial nerve deep in the brain. The compression was usually caused by an impinging blood vessel but sometimes could also be caused by a tumor. I finally had a name and a plausible explanation. I praised Marty, informed him that I had an appointment with my primary care physician and would discuss it with him. Before going to my doctor, I read up on the condition. I learned that pain medications were ineffective and that surgery could be done to decompress the nerve. In some cases, where decompression was not successful, severing of the nerve was an option although this wasn't desirable as it often causes drooping of the face.

When I got in to see my doctor, I started to fill him in. He was already aware of the ENT's diagnosis and the fact that I was seeing a psychiatrist. I went on to explain that I had discussed my symptoms with an internist at work and that he thought glossopharyngeal neuralgia was a possibility. My doctor said, "So what if it is. There's nothing you can do about it. You just have to learn to live with it. Are you still taking your Paxil?" I told him that there was something that could be done about it. I said, "The nerve can be surgically decompressed or severed. I even read about an experimental noninvasive surgical technique using a device called a gamma knife to decompress the nerve." He responded by telling me that glossopharyngeal neuralgia often self resolves and indicated that I would be better off to avoid risky neurosurgery. He then went on to say that I needed to focus on becoming emotionally well first. I could see that I was getting nowhere, thanked him for his time and left.

It was now mid June. I had had it with Paxil, the psychiatrist, and the good intentioned but ineffective therapist. I called Ramona and complained about the care I was receiving. She was very receptive to my concerns and said she had just the right psychiatrist for me if I was willing to go down town to the Jewish Hospital Medical Arts Center. I told her I would try anything to get better. Ramona then referred me to Dr. Marilyn Sholiton. She said that it was difficult to get in to see Dr. Sholiton but she would make a call for me. Ramona did her thing and got me an appointment with Dr. Sholiton for the following week. I stopped taking Paxil cold turkey. My abdominal pains went away in a few days but the insomnia didn't improve. When I met Dr. Sholiton, I liked her immediately. She was in her fifties and I could tell she was Jewish. As a child, I lived in a predominantly Jewish neighborhood. There was a certain caring quality that Jewish mothers had and Dr. Sholiton possessed that quality. She took me back into her office and started by telling me that each person was a book and asked me to start telling my story. As I told my story, I cried once again. I also talked about my bad experience with the previous psychiatrist and my unsuccessful attempt to take Paxil. She wanted to treat me for both anxiety and depression. For anxiety, she wanted to continue using Ativan and add BuSpar, for depression, Prozac. She asked me for my permission, which I found odd. I told her I was skeptical of Prozac because it was an SSRI like Paxil. She explained that each person reacted differently to each antidepressant and that, in her experience, people often found success when switching from one SSRI to another. The only way to find out was to try and if it didn't work, we would try something different. I agreed to try the therapy and she said that she would also be pleased to be my psychotherapist.

Somehow before I left, we got on to the topic of religion. I must have started by telling her that I was seeing a priest. I could see it bothered her. She asked me to share what we talked about and I agreed. I asked her if she had a problem with Catholicism. She said, "no, why?" I said, "I don't know, maybe because you are Jewish." She asked how I knew she was Jewish. I said that this was the Jewish Medical Center and that I had been raised in a Jewish neighborhood in Philadelphia and she reminded me of the neighbors I was so fond of. She confirmed that she was indeed Jewish but that it had nothing to do with her concern about me seeing Father

Al. She went on to explain that she didn't want her patients to think that they were suffering from depression because God was punishing them. I assured her that that was not the case. Before I left, she remarked about my first name, Raphael. She said it was Jewish. I told her that indeed it was. Raphael was the ancient Jewish word that meant, "God heals", and that in Catholicism, St. Raphael was the patron saint of healing and doctors. I told her that I had been named in honor of St. Raphael and that St. Raphael could be found in the Catholic Bible but not in either the Jewish or St. James Bibles. Our relationship had gotten off to a great start and little did I know the critical part she was about to play in my life.

At my second session with Dr. Sholiton, she indicated that she wanted to rule out anything physical being wrong. She did not know the ENT I had seen and wanted me to see an ENT she knew and trusted at the university. The second ENT was also very personable. He examined me by pressing along certain nerve routes on the inside of my mouth. He said that if I had glossopharyngeal neuralgia I would not be able to stand the pressure he was applying with his finger. I asked him if it could be a tumor because my older brother had a brain tumor. He said no. He then took my hand and said, "I am telling you man to man that you do not have a brain tumor and that I will be here to see you through this thing." It was his polite way of telling me that I was depressed and he would support me. He went to say, "We can do an MRI scan of the brain but it is expensive and it will only be negative." I thanked him and left.

I now had two opinions from two different ENTs that nothing was wrong. I began to question my resolve but kept coming back to the same thoughts. Why does the right side of my face feel normal? Why would depression discriminate against my left side?

My reaction to BuSpar and Prozac was not good. The GI pain returned with even greater intensity but I continued to take it for a few weeks. My insomnia also continued to worsen. I was going 2 or 3 days without sleep routinely. When I did manage to sleep I would only get 4 hours at most. I never found myself sleepy and found it impossible to nap despite numerous attempts to do so. I drew a distinction between being sleepy and feeling tired. While I was tired beyond belief, it didn't make sense that I never had headaches. In the past, if I missed a night's sleep due to international travel for example, my head would throb. I complained to Dr. Sholiton about both the GI pain and insomnia. She referred me to a gastroenterologist because GI side effects from Prozac were usually not as bad as I was describing them. For insomnia, she prescribed a particularly effective sedative called Restoril.

That night, I took 15mg of Restoril and for the first time in months slept a straight 8 hours. The very next night it didn't work. The next night I took 30 mg but again it didn't work. The gastroenterologist she referred me to was someone I actually knew. She lived next door to us 14 years ago before we relocated to upstate New York. She ordered a series of tests that included a CAT scan, ultrasound, and lower bowel x-rays. All the tests came back negative. To help me with sleep, Dr. Sholiton discontinued the Restoril and added Klonopin, another sedative, with my permission of course. She told me to take the Klonopin and get in bed. Once in bed, I was then to place two Ativan tablets under my tongue. She said I would fall asleep in 5 minutes. The Ativan would be directly absorbed into my blood stream from under my tongue. This is known as sublingual administration. The Ativan would put me to sleep and the Klonopin would keep me asleep, as it required more time to get absorbed from the stomach and into the blood stream. I was also to discontinue BuSpar, as the Klonopin would also better help manage my anxiety. I tried the new cocktail that night but didn't find it very effective.

Since Prozac didn't seem to be improving my mood after a few weeks, Dr. Sholiton took me off it. The intense GI pain gradually diminished over a period of a week or so; consistent with the fact that Prozac stays in the body after discontinuation for several weeks. The next problem was what antidepressant to try next. We considered those known to be strongly sedating, most belonged to a class known as tricyclics. The tricyclics had been displaced in the market when the newer, better tolerated, SSRIs, like Prozac, Paxil, and Zoloft, came along. I had read about the tricyclics. In addition to drowsiness, they tended to cause dry mouth, weight gain and constipation. I inquired about a new antidepressant, Remeron, that was neither a tricyclic or SSRI. She was aware of it but had not yet prescribed it to any of her patients. Remeron was also supposed to be sedating because, in part, it possessed antihistamine properties in addition to its antidepressant activity. We decided to try it. I would be her first patient to take it. She suggested that I take it after dinner.

That evening after dinner, I took 15 mg of Remeron and went into my study to work. About 45 minutes after taking the drug, I felt pressure around my forearms like someone had wrapped my arms with blood pressure cuffs and pumped them full of air. Shortly after that, I began to feel very drowsy. I was elated that we found something that worked. I immediately headed up to bed and slept through the night. The Remeron was so effective over subsequent nights that I tapered down and went off the Ativan in about one week. The pressure sensation in my arms went away but was replaced with the sensation of a mass inside my stomach that was pushing outwards. It wasn't painful, just unusual. By now, it was August. Since the Remeron was working so well, we made plans to visit my parents in Philadelphia before Dominic had to return to school at the end of the month.

While at my parents, I was very quiet, depressed and internally obsessed with my symptoms – the facial pain and numbness, the choking around the neck, weakness, nausea, frequent urination, daily diarrhea, a heightened sensitivity to noise, light and odors, tingling in my left hand and foot. At least the Remeron was letting me sleep which made an enormous difference in my ability to cope. My parents were aware that I was suffering from depression but weren't sure how to talk with me. My father pretty much left me alone but my mother kept repeating, "You look fine. What's wrong? You can snap out of it." The funny thing was that I could understand her confusion. After all, how could someone look fine and have such a long list of complaints? The vacation was beneficial as I escaped the demands of work for a week.

During the month of September, I went to Dr. Sholiton twice a week for therapy. We covered a great deal of ground. How I was raised. My relationships with my parents and brothers. My education achievements. My religious beliefs and discussions with Father Al. My career and the workplace. Sometimes we would talk about science or technology or medical advances. The medications were giving me vivid dreams when I managed to sleep so we talked about the ones I could recall. I often dreamt of being trapped by fire or traffic jams or crowds. She interpreted the dreams as the feelings I was having about my career growth, which had slowed over the recent years. I pushed back and said that the dreams were a reflection of my frustration with the medical system and lack of a proper diagnosis. I eventually indicated that I didn't know where our discussions were going to which she replied, "You'll talk to me when you're ready." When a new symptom emerged, my anxiety level would increase and I would again express my concern that I had a brain tumor. Every time this would happen she would rationalize with me, which was helpful in calming me down. She would say, "You don't have headaches, right? Your pupils are the same size, right? You don't have vision problems, right?" Her logic was solid to the point where I began questioning myself. "Perhaps depression can cause all of these physical symptoms. Perhaps depression is causing some of the symptoms and the medication is causing some."

The 15 mg of Remeron, which had worked so well for sleep for about a month, began to lose its effect. The dosage was adjusted to 22.5 mg which worked for a few days then it was increased again to 30 mg but its sleep promoting effect only lasted a couple of days. It was clear that I was adjusting to the medication. At the same time, we tried more Klonopin, going as high as 2 mg. After a couple of weeks, it looked like insomnia was back to stay. Dr. Sholiton explained that difficulty falling asleep was due to anxiety while difficulty staying asleep or waking early was a sign of depression. I had both in spades. My mood was not improving very much either. It was better on days when I managed to get some sleep the night before. As psychotherapy progressed, Dr. Sholiton learned that I was a perfectionist and had a tendency to focus on a particular task until it was completed. She felt that I was obsessed which I couldn't deny. I was obsessed with my symptoms. I remained fixed on the notion that the doctors had missed something terribly wrong and I was constantly worrying about not getting enough sleep. She went over all the right practices for dealing with insomnia. Don't take naps. Don't lie in bed awake. Etc. None of it worked. We began to consider identifying a hobby or task that I could engage in to focus my thinking away from my problems. I had read a book by Dale Carnegie entitled "How to Stop Worrying and Start Living". In the book, he told a story of a depressed man who made a list of everything that needed repair around the house. The man not only worked his way out of depression but substantially improved the quality of his home as well. I tried it. My house was fairly new but I managed to find a number of things in need of repair. I found doing the work to be helpful but there just wasn't enough of it. I needed something else to do.

At the next session, I offered art as an idea. When I was in my teens, I liked to draw but I always wanted to try painting so I decided that I would teach myself how to oil paint. I chose to start with the wet on wet technique, an easy method of creating land and seascapes. I began painting every night after dinner. I found that it required an incredible amount of concentration. I also found it stressful but in a good way as I wasn't thinking about symptoms or anything else other than trying to create a good painting. Over a couple of weeks, I began to sleep better and my mood began to shift in a positive direction. I painted every chance I could get. It wasn't unusual for me to complete as many as five paintings in a week. It also turned out that I had a modest amount of talent. All of my relatives and in-laws wanted my paintings. I painted scenes for Dr. Sholiton, her receptionist, and Father Al. Eventually, some of my paintings would be successfully auctioned in a fund-raiser for a local mental health organization. While the character in Carnegie's book worked his way out of depression, I was painting my way out.

By November, my depression and anxiety began to lift significantly. I was once again free to experience the every day common joys of life. My concentration returned to some extent and I was better able to perform at work and reclaim many of the responsibilities that I had delegated to my subordinate managers. Sleep improved a little but it was far from acceptable. There continued to be nights every week when I didn't sleep at all and I never slept longer than four hours. Dr. Sholiton was certain that once the depression lifted, my physical symptoms would fade away. After all, the associate she dearly trusted agreed that I was merely depressed. She inquired as to the pain in my face. I informed her that there was no change. All of my physical symptoms remained plus my right hand now began to tingle. I told her that my arms continued to weaken. It was fall and I had recently needed to remove the leaves in my gutters. I found that I was unable to lift my extension ladder without the assistance of my son. In the past, it had never been a problem to lift and carry the ladder alone. As my major complaint continued to be insomnia, Dr. Sholiton and I kept tinkering around with doses. She had heard from a sales representative that Remeron was energizing at higher doses and sedating at lower doses. This could not be confirmed in the FDA approved labeling of the drug nor could it be found in any scientific publications. The highest dose Remeron was approved for by the FDA was 45 mg. In Europe it was approved for as high as 60 mg. We settled in on a dose of 30 mg, which I maintained throughout most of the winter.

Things remained fairly stable over the winter in terms of my physical symptoms. I continued to paint and take my Remeron and my mood continued to improve. I told Dr. Sholiton that coming downtown twice a week was getting inconvenient and that I wanted to reduce my sessions to once a week. She agreed and felt that I had made significant progress. In late winter, my insomnia worsened. I went up to 45 mg of Remeron, which helped for a few nights. Then on the theory that my depression was really being driven by obsessive-compulsive disorder (OCD) and OCD usually required higher doses of antidepressants, we went all the way to 60 mg. After a couple of weeks at the higher dose, I began to get severe debilitating headaches. I wasn't sure if the headaches were from the increased dose of Remeron or just part of my degenerating condition. My legs began to get weak and walking became difficult. If I attempted to walk at a normal pace, I would get sick to my stomach. I would also get dizzy when walking down hallways. And, I just felt sick all of the time. The time had come for action. I grabbed my buddy Marty and filled him in on all my symptoms and medications. Marty still believed I had a problem with my posterior fossa (the hindbrain). A year earlier, when I was on the Internet investigating glossopharyngeal neuralgia, I came across the Mayfield Clinic website where Dr. John Tew was listed as an expert in cranial nerve decompression. I decided to go straight to him so I called his office and made an appointment for two weeks out. I couldn't get in to see him sooner as he only saw new patients on Tuesday afternoons.

Chapter 5
MRI Reveals the Truth?

It was now April of 1999. I was a mess physically. I was fatigued, weak, and nauseous. I was urinating frequently and had diarrhea daily. There was the pain on the left side of my face and more recently, extreme headaches. It constantly felt like someone was choking me to the point where my head was going to explode. My hands and feet were constantly tingling. Dr. Tew was an older gentleman. He was head of the department of neurosurgery and a former president of the American College of Neurosurgeons. Oddly enough he tried to save my wife's younger sister 20 years ago when an aneurysm burst in her head at the age of twelve. He was quiet by nature but very confident. I described my symptoms to him, informed him that I was taking medication for depression, and expressed concern about a possible tumor. He asked me more questions and performed a neurological examination. Following the exam, he said that I was probably having tension headaches but it would be wise to get an MRI of my head. He stated that he honestly didn't recognize the constellation of symptoms I was having but that he had been around long enough to know that he could be fooled from time to time. Ordering an MRI was music to my ears. I didn't care if he was being honest with me or if he was just doing it because he appreciated that anxious, depressed patients sometimes need proof to relieve their worries.

I underwent the MRI scan about a week later. It was performed with and without contrast agent in order to detect tumors. Other than the jackhammer like noise of the machine, it was rather uneventful for me. Apparently, some people find it very difficult to be closed up in the instrument's tight tunnel. A couple of days later, while at work, I received a phone call from Dr. Tew's nurse, Nancy. She said that they found out what was wrong with me. I had something called an Arnold-Chiari malformation of the brain and that I shouldn't worry because they had a surgical procedure to deal with it. I asked her to explain it to me. She said that it was a rare birth defect of the brain that caused crowding where the skull meets the spine. She said that the surgery consisted of removing some bone and making room for the defect. I was then asked to return to the office in two days so that Dr. Tew could answer my questions and talk to me about surgery.

As I hung up, I was overwhelmed with emotion. I was finally vindicated. There was something wrong with my brain. Marty and I had been right. I immediately turned to the Internet to see what I could find on Arnold-Chiari malformation. The first hit was a site by Chip Vierow entitled "Chip's Arnold-Chiari Malformation Page". I read it with great interest. I learned some of the basics. An Arnold-Chiari malformation was a congenital birth defect of the brain where the cerebellar tonsils at the base of the brain are elongated and descend down into the upper spinal canal. This condition caused cerebral spinal fluid to drain poorly from the head resulting in increased pressure to the cerebellum, brain stem and lower cranial nerves. A German pathologist by the name of Hans von Chiari first observed the defect during autopsies in 1891. Later, another German pathologist, Julius Arnold, described his pathological findings of the condition. The malformation was classified as types I through IV depending on the parts of the brain displaced. Type I was the most common and usually the type observed in adults. The modern term used was just Chiari Malformation Type I or CMI. Chip described his symptoms, which were so amazing similar to my own. His site described the surgery, known as decompression surgery, and even included photographs. I learned that the surgery was more involved than what Nancy had just described to me. It not only consisted of removing bone from the skull and upper spine but also required opening of the Dura and the removal of adhesions on the cerebellar tonsils and brain stem by microsurgery. Finally, the Dura was enlarged with a bovine pericardium graft.

The procedure described and shown in Chip's web site didn't really phase me. If it could cure my symptoms, I was game. From Chip's site, I found the World Arnold-Chiari Malformation Association or WACMA web site. On that site, I found an unfiltered survey list of symptoms as reported by 31 diagnosed individuals. The information was interesting. I did not have some of the more common symptoms such as loss of vision and general imbalance. But, I did present with many of the other symptoms on the list. I was keenly interested in insomnia and discovered that it was reported on the list by 61% of the respondents. Later I would learn that insomnia included all reasons such as insomnia due to pain and discomfort, and insomnia from central sleep apnea. There was also a list of recommended doctors, where I was pleased to find Dr. Tew's name. I printed out much of the material to take with me to my next appointment with Dr. Tew. I next called my wife to share the good news. She was surprised to some extent, as over the last year, the doctors had pretty much convinced her that my symptoms were psychogenic in origin. Father Al was on my calendar before I was due to return to Dr. Tew's office. I told Father Al that they found a brain defect and that I would be having major brain surgery. At the end of my session with Father Al, we walked a couple of blocks across campus to the chapel where he administered the sacrament of the anointing of the sick on me.

When I returned to Dr. Tew's office, I was pumped and anxious to discuss surgery. I was no longer having headaches because I decreased my Remeron from 60 mg to 45 mg. Apparently, the headaches that Dr. Tew classified as tension headaches were being caused by the high dose of Remeron I was taking. After waiting a few minutes in one of the examination rooms, a young doctor entered. He introduced himself as Dr. Tew's chief resident. I told him that I was expecting to talk with Dr. Tew as Nancy indicated to me on the phone that Dr. Tew wanted to discuss surgery with me. The young doctor smiled and said that surgery wasn't necessary. He explained that he had reviewed my scans and really didn't consider me to have Chiari because the amount by which my cerebellar tonsils descended below my skull was insignificant. Further, in his opinion, there appeared to be sufficient space for CSF flow and thus my symptoms couldn't be caused by the malformation. I asked how far my tonsils were herniated. He said that he didn't know, picked up one of the films, held it up to the light and said about 5 mm. He said that people with Chiari had to have herniations of at least 10 mm. I handed him the information I obtained from the Internet and told him that I shared many of the known symptoms and that early surgery was recommended. He seemed surprised and asked me where I obtained the information. I told him that I found it on the Internet. He commented that you have to be careful about the medical information one obtains from the web. Dr. Tew then walked into the room and asked me if his resident explained their opinion to me. He added that surgery was not necessary as my symptoms were mild and the surgery wasn't risk free. He said that Chiari usually presents with headaches in the back of the head and intense burning in the arms and legs. I had no way to judge the benefit to risk ratio. All I knew was that I was very sick and wondered how sick does one have to become before someone will help you. Dr. Tew asked me if I was still seeing my psychiatrist. I told him yes, he said good and left. I asked his resident what was next. He explained that there was a more definitive MRI test that they could do called a Cine MRI but that he recommended we start with it next time if my symptoms were to worsen in the future. As I left the office, a group of doctors were standing around looking at the material on Chiari that I brought, shaking their heads and exclaiming their amazement as to how such misleading information could be out there on the net.

As far as psychiatric care was concerned, I was seeing Dr. Sholiton on a biweekly basis. I was no longer depressed. She was also sure I was no longer depressed but was confident I had OCD. We talked about OCD. It wasn't necessarily a bad thing. If properly focused, it helped me achieve important and productive goals. She felt however that it wasn't properly focused. I naturally disagreed. I thought being focused against the goal of attaining good health was very proper. A lot of our time was spent talking about medication and dose adjustments or the topic of insomnia. I filled her in on Chiari and my experiences with Dr. Tew. Over the next few months, we would talk a great deal about Chiari as I learned more about it.

I returned to the Internet to learn more about Chiari and found the WACMA on-line support group as well as a group known as the American Syringomyelia Alliance Project or ASAP. Syringomyelia is the medical term for a spinal cyst, which is caused in most instances by Chiari. ASAP is a non-profit organization dedicated to finding a cure for both Syringomyelia and Chiari. The on-line support group was a gold mine of information. There I was able to talk with many people who had been diagnosed with Chiari and surgically decompressed. One of the first things I learned was that no two Chiarians present in the same way. No one has the same malformation anatomy or symptoms. When I asked the group how long tonsillar herniation must be to be considered Chiari, I learned that 10 mm was an old guideline and that the new cutoff was 5 mm. Further, I learned that the length of herniation wasn't necessarily the important factor. What was important was the overall shape of the malformation and how it was crowding the opening at the base of the skull also known as the foramen magnum. It was also important in the case of borderline herniations to confirm CSF blockage by Cine MRI. Cine MRI allows the radiologist to follow the actual flow of CSF as it drains from the skull into the spinal canal through the foramen magnum. If the malformation is blocking or impeding the flow, it can be readily detected and even measured quantitatively by Cine MRI. To educate myself further on Cine MRI, I rented a videotape on the topic by a National Institute of Health neurosurgeon from ASAP's videotape library. After viewing the tape, I realized that Dr. Tew's resident should have proceeded with a Cine MRI and not have assumed that sufficient space in the foramen magnum was present. Nevertheless, I didn't know if a positive finding by Cine MRI would really change anything as they were of the opinion that my symptoms didn't yet warrant surgery.

The following months were the most difficult in my life. I was growing weaker. My arms were so weak that I would place my forearms on my lap for support when driving the car. I was sick to my stomach and felt like I had the flu all the time. Walking was becoming difficult. I couldn't mow the entire lawn by myself. Fortunately, my son was always there to finish the job. When I did mow the lawn, I would recite the rosary to myself and plead with God for help while thinking of Psalm 18. The group of physicians I ate lunch with at work routinely enjoyed taking a walk around the research center every day after lunch. It was becoming difficult to keep up with them. If I exerted myself, I would get over heated and feverish. We had a blue spruce in our front yard that had died and needed to be taken down. I asked my son to do it. When he began to get tired, I stepped in and swung the axe a few times. I got so over heated that I went in the house, filled the bath with cold water and soaked in it for 45 minutes. I was sick the rest of the day and couldn't sleep at all that night. I had a similar experience just from using my arm to scour a dirty frying pan. To make matters worse, I was general chair for the American Chemical Society's (ACS) Central Region Meeting to be held in June of 2000. The conference was fast approaching and the committee and I were very busy. It was an additional strain I didn't need. I told the committee that I was diagnosed with a rare brain condition and that I might need to step aside in the future if my condition worsened. One of the members, Kathy Gibboney, called me at work a couple of weeks later. She told me that she had a personal friend that she had grown up with by the name of Mary Kay Gummerlock, who was a neurosurgeon in Oklahoma. They were recently talking and Kathy's friend commented that one of her special interests was Chiari. Kathy proceeded to tell her about me and she indicated that she would be very interested in reading my MRI scans and providing a second opinion. I welcomed the opportunity and Kathy provided me with her address and phone number.

I immediately sent my MRI scans to Dr. Gummerlock. About a week later, Dr. Gummerlock called me at my home one evening. She said my malformation appeared minor but couldn't rule out that it may be causing me problems. She qualified that everything she would say regarding an opinion was to be taken as tentative since she didn't have the advantage of seeing me in person. She proceeded to ask me about my symptoms. I also told her that I was on medication for depression and that my doctors believed my symptoms were psychogenic in nature. After speaking with me for about an hour, she concluded that a Cine MRI was indicated. I told her that I couldn't very well go back to Dr. Tew and tell him that you think he is wrong and that he should order a Cine MRI. She agreed and suggested that I make an appointment with my primary care physician. She then went one step further and indicated that she would call him prior to my appointment and educate him on Chiari and what should be done for me. Once I had evidence of a problem by Cine MRI, I could return to Dr. Tew for help. I was amazed that a stranger was so willing to help me. I thanked her multiple times.

Another week passed, it was now August. I got in to see my primary care physician and told him of all my symptoms. He listened attentively and gave me a physical exam. Not knowing if he had yet talked with Dr. Gummerlock, I brought up the suggestion of getting a Cine MRI. At that point, he revealed that Dr. Gummerlock had called him earlier in the day and that they had had a long talk. He still seemed very resistant on the idea of ordering a Cine MRI and asked if I was still seeing my psychiatrist. I told him that I was and that he should call her because she could confirm that I was no longer mentally depressed. I then said, "Maybe I sound crazy. I don't know. But, you have known me for several years and should know that I am not the type who wants to be ill." He then responded by saying, "I have to tell you. The whole thing sounds crazy to me. If you do have Chiari, you'll be the first patient I ever had with it." I continued to press for the MRI. He then said that he would do his best to order it but warned me that my health insurance may not approve it. I told him, that if necessary, I would pay out of pocket for it. I knew he had no intention of helping me. He didn't believe I had Chiari. Being in the health care industry, I was also aware that many doctors had contracts with the health insurance plans which provided them with a financial incentive for keeping the cost of expensive diagnostic tests down. In essence, if they keep the cost of diagnostic tests under a certain per patient amount, they got a bonus at the end of the year. I couldn't help but wonder if he had such a contract. After a few days, my doctor's nurse called me and informed me that my insurance would not approve the Cine MRI. I then reminded her that I would pay for it. At that point, she simply said that the doctor thinks I should return to Dr. Tew for it. As far as I was concerned, I had just wasted another two weeks and time was running out.

I called Dr. Gummerlock to tell her of my failed attempt and obtain her advice on what to do next. She encouraged me to go back to Dr. Tew and expressed confidence that he would now help me. When I called Dr. Tew's office to make an appointment, the soonest they could get me in was in seven weeks. I told the receptionist that I was really sick and needed to see him sooner but my plea fell on deaf ears. I didn't think they were putting me off on purpose. I knew patients had trouble getting in to see my brother who was a top pulmonologist at a major university hospital as he was often on the road making presentations to other doctors at various medical conferences and pharmaceutical companies' special events. I figured Dr. Tew also had a busy surgery schedule in addition to similar commitments.

It was mid August and my son was to be awarded the rank of Eagle Scout at a special outdoor ceremony at the town square. I had been looking forward to the event all summer. I didn't want Chiari to take me out before the ceremony. As I stood by him, my legs nearly buckled. Remaining on my feet for a couple of hours afterwards to socialize with the other parents was agonizing. Dr. Sholiton even attended. We had grown very close over the past year and she was interested in meeting my son. I struggled the following two weeks at work to stay on my feet. I didn't walk with my friends at lunch. I came home one day in late August, went upstairs to change and collapsed.

I called out for my wife. Marilyn came up stairs immediately to see what was wrong. I was lying on the bed. I told her that I no longer had enough strength in my legs to walk to any extent. The following day I called my boss and told him that I was too sick to come in and that I didn't know when I would be in again. I told him that I was trying to get in to see Dr. Tew but that I was having trouble. I asked if I could continue to work from home by email and phone until I could get in to see the doctor. He agreed. Next, I called Dr. Tew's office and explained what had happened. His office insisted that nothing could be done to move my appointment up. As I had an appointment coming up with Dr. Sholiton, I called her office next to cancel it. She became very concerned and called Dr. Tew's office to clearly state her position that my condition was not psychosomatic. Shortly after that, Dr. Tew's office called and moved my appointment up five weeks. I was to come to his office in two weeks.

I set up shop in the family room. I lay on the couch all day with my laptop and phone by my side. I would only get up to go to the bathroom and eat my meals. My subordinates called in for their conferences with me. My boss offered to come to my house to meet with me once a week. I told him that I needed to get out once in a while and suggested we meet for lunch once a week. My wife would drive me to the front door of the designated restaurant and I would shuffle my way to the table. I called the vice chair of the ACS meeting committee and asked him to take over for me. All I had to do now was wait it out.

Two weeks passed by. It was now September and the time had arrived to return to Dr. Tew's office. I didn't know what to expect. My wife drove me to the office and let me off at the nearest entrance. I shuffled to the waiting room and then back to one of the exam rooms when called. After a few minutes, one of Dr. Tew's associates came into the room and told me Dr. Tew would be in shortly. He said, "You're the gentleman we're following for Chiari." He then gave me a neurological exam. As usual I passed it. Dr. Tew with about 3 other doctors then came in the room. I told him how sick and weak I felt and that my arms and legs were now burning like fire. He said in a soft voice that the time had come to do surgery. The other doctors took a great interest in me and began to ask me questions about my symptoms comparing me to other

Chiarians that they cared for in the past. One asked me if I felt a metallic pinging inside my chest. I told him no but that I felt it in my throat. I also told them that I could rub my hand over my right shoulder and feel it in my throat. I told them about many of the odd feelings I was experiencing like having a tongue that was numb on one side, individual teeth that were in pain, the choking around my neck, the hypersensitivity to light, sound and odor, and a great deal of crackling that I could hear in my head when I rotated my neck to any extent. They were all taking notes. I felt like a celebrity and wondered why the reception was so different this time. Dr. Tew then gave instructions to schedule me for surgery and get a Cine MRI on Saturday just to confirm the diagnosis. One of his associates then asked if an MRI of my spine should be taken to look for Syringomyelia. Dr. Tew said no and explained that my symptoms were consistent with brain stem compression and not a syrinx. He went on to say that surgery was now imperative in order to prevent Syringomyelia. We even joked around a little. I asked Dr. Tew about the safety of the bovine pericardium tissue. I told him I didn't want to wake up with mad cow's disease. He asked if I would prefer him to use my own pericardium tissue. My long ordeal was over. Help had finally arrived.

I went in for the Cine MRI on Saturday and back to Dr. Tew's office the following Tuesday. This time, a doctor I did not see before came in to talk to me. I got another neurological exam, which, naturally, I passed. I told him that the neurological exam was probably a hundred years old and lacked the sensitivity to confirm many

Chiari symptoms. To understand why I passed the neurological exams, it is important to realize that weakness, for example, can be both objective and subjective. I could demonstrate strength when objectively tested for it. My weakness was subjective in nature. When I attempted to use my strength, even to the slightest degree, I would become extremely ill. As a result, I had to hold back on everything I did. While the neurological examine didn't confirm my illness, I was told that the Cine MRI did. I asked if I could see the scans. He said, "Yes, but there is nothing to see. You are completely blocked." He then patted me on the thigh and said, "We owe you the best surgery we can give you." I interpreted the remark as the closest thing to an apology that I was going to get. He then asked me, if he could wave a magic wand over my head and make three symptoms disappear, what would they be. I told him insomnia, fatigue and facial pain. Then I got the bad news. It was made clear to me that they couldn't promise improvement with the surgery. The goal of the surgery was to arrest further decline. If improvement was obtained, it would be considered a bonus. I told him that I understood but needed the surgery desperately. From my reading, I had an understanding that about 20% got significant improvement, about 70 to 75% got some improvement, and 5 to 10% got no improvement or worsened. After a discussion on the risks of the surgery, I signed an informed consent form. The primary risks were infection, paralysis, a dural leak, and even a small risk of death as with any operation that required general anesthesia. He then told me that they would try to schedule my surgery for the next Tuesday early in the morning but if they couldn't fit me in then, I would have to wait three weeks because of Dr. Tew's travel schedule. He ended by asking if I had any questions about the procedure. I told him no because I had done a lot of reading about it and spoken over the Internet to many people who had had the operation.

I called Dr. Gummerlock one more time to tell her of the outcome and also to thank her once again as I suspected that she might have even called Dr. Tew. She was interested and asked me a couple of questions. How long did they tell you, that you would be in the hospital? I told her three days. She replied that it would probably be a little longer. She then asked, "How long did they tell you that you would be off from work?" I told her 3 to 4 weeks. She said that it was major surgery and I would need between 6 to 8 weeks to recover. She wished me luck and we hung up.

Dr. Tew's office called me the following day and told me that I would have to wait three weeks. They tried everything they could to fit me into his schedule but it just wasn't possible. I was disappointed but I had waited a year and a half so I figured I could wait three more weeks. The date for surgery was set for Friday, October 8, 1999. I was to report to the hospital at 5 a.m.

The next stop was my primary care physician, as I was required to get a preadmission physical. My doctor performed the physical and said that I was now in the hands of the big boys and that they would fix me up.

He would make no effort to inquire on my status while in the hospital and I would never return to his office again.

I asked my boss if I could continue to work from home until the surgery, as I wanted to keep my mind occupied. He readily agreed and so the wait began. For three weeks, I worked and waited on my back. Friends and coworkers called me as the news began to leak out. Of course, no one had heard of Chiari so I found myself doing a great deal of explaining. The most popular question was. "If it's a birth defect, why does it take so long to become symptomatic?" That was indeed the question and one that I asked myself to Dr. Gummerlock. She informed me that no one really knew but that there were two schools of thought. The first was related to arthritic changes from aging at the skull-spine junction. The second had to do with the growth of adhesions over time as a result of the cerebral tonsils trying to stabilize themselves against the motion they were subjected to in the upper spine. When I asked Dr. Tew this question, he said he wasn't sure but it may be due to the skull settling on the spine upon aging causing the tonsils to descend a few millimeters further into the spinal canal and blocking the flow of CSF. Since the average age of symptoms emerging is only 25, explanations based on aging seem less probable. Thus, the formation of adhesions seems to fit the situation the best.

Of course, the second most frequently asked question was, "What does the operation entail?" The basic idea behind the procedure is rather simple – to eliminate the pressure on the brain by making room at the craniocervical junction. The term for the procedure is decompression. Some more specific descriptive variants include, decompression of the craniocervical junction, decompression of the foramen magnum (large opening at the base of the skull), and decompression of the hindbrain. The decompression is achieved by first making an incision that extends from the base of the back of the neck to about an inch from the crown of the head to expose the craniocervical junction. A small amount of bone about the size of a half-dollar from the base of skull is removed along with the posterior arches of the first and sometimes the second cervical vertebra. The dura, which is the tough elastic membrane that surrounds the brain and spinal cord, is then opened to expose the herniated cerebral tonsils. With the aid of a special microscope, adhesions between the tonsils and the brainstem and upper spinal cord are carefully removed. Some surgeons prefer to remove the tonsils altogether since they have no known function. A graft is then placed in the dura to expand it. A number of different types of graft synthetic materials or natural tissues can be used. One of the more common ones, and the type that I would receive, is bovine pericardium tissue, the tough elastic tissue that surrounds the heart of a cow. (Note: Apparently, there is some concern about using bovine tissue because of mad cow disease. Despite the special processing bovine tissue undergoes, patients with bovine grafts can never donate their blood.) Removing bone and adhesions, and expanding the dura creates the necessary room to return the flow of cerebral spinal fluid back to normal. By returning CSF flow back to normal, the compression to the cerebellum, brainstem and lower cranial nerves is eliminated.

The idea of undergoing this procedure did not bother me. I was so sick that I was willing to try just about anything. I wondered if getting surgery locally was the right decision. While my surgeon was renowned and had even published papers in the medical literature on Chiari, there were a handful of surgeons in the country who specialized in Chiari decompression surgery. Many Chiarians at the on-line support group advocated going to these specialists but I was too ill to travel and lining surgery up away from home would require additional time. I also figured that a positive to having the surgery locally was a greater ease of getting follow up care.

By this time, my mental state was sound. I was resolved to get better. I was still on medication, taking low doses of both Remeron and Klonopin, which helped me get four hours of sleep a night from about midnight to 4 a.m. In addition to work, I kept myself busy by doing crossword puzzles. During this period, my younger brother, Gary, came to town on business and took the opportunity to visit. My walking was restricted to going to the bathroom and the kitchen table for meals and I had acquired a characteristic shuffle where I hunched over and took very small steps without lifting my feet off the floor. Walking in that manner seemed to help minimize the nausea. When my brother observed me shuffle to the table for dinner, he became furious and remarked, "I can't believe those damned doctors let you get this sick." I was also beginning to have difficulty breathing particularly early in the morning when I would first awake. One Sunday morning, it was so bad I called Dr. Tew at home and asked if the surgery could be moved up even if another surgeon had to do it. I caught him on his way out the door to catch his flight out of town. He quickly assured me that Chiari wasn't fatal and convinced me to wait it out until he returned.

The day before surgery, my parents arrived for support. I wasn't able to sleep at all that night. My wife got up early and we departed for the hospital at 4:30 a.m. Since my parents were from out of town, a neighbor volunteered to take them to the hospital by 7 a.m., the time my surgery was schedule to begin. We arrived at the hospital just before 5 a.m. Before I was prepped for surgery, I was required to complete a number of forms. Some had to due with insurance and some wanted such information as how tall I was and what medications I was taking. I had provided all of this information to the hospital earlier that week but they had their system. A nurse then took me to a small room where I was asked to change into a hospital gown and get on the bed. An i.v. line was inserted into my left arm and, by protocol, I was given a small amount of sedative to keep me calm. From there, I was wheeled into the recovery room where I was placed next to a moaning patient coming out of anesthesia. While there, the anesthesiologist inserted an arterial line into my right wrist. Later I heard voices of disapproval in the background. Apparently, I wasn't supposed to be in the recovery room and someone was trying to explain why he or she wasn't able to take me to the room I was supposed to be in. I wasn't sure but it seemed that someone was concerned that the patients coming out of anesthesia might upset me. Whatever the concern was, I was perfectly comfortable.

I waited awhile in the recovery room before being wheeled into the operating room. Without my glasses, I had trouble seeing where I was and who exactly was in the operating room. I didn't see or hear Dr. Tew but figured he wouldn't come in until I was fully prepped and put under. I was lifted off the bed onto a table. My gown was removed and leads were placed at different locations on my body. Someone explained that they would be sending small shocks to my different muscles during the operation to make sure they didn't damage any nerves. One of the physicians then walked me through what amounted to a verbal consent. He asked me why I was there. I told him I was there for a decompression of the foramen magnum. He then asked me if I could tell him some of the risks of the operation. I said something like infections and dural leaks to which he replied, "and of course, you realize that there is a small chance of death from general anesthesia." I said yes, raised my right arm and gave him a thumb up. A mask was then placed over my face and I was instructed to count to four. I made it to four and feel into a deep sleep.

The next thing I knew, I was freezing cold and my body was thrashing about. I could hear voices but couldn't wake up. People were holding me down and calling for more warm towels. A comment was made that I was bucking like a wild bronco. Someone else mentioned something about another patient who had an allergic reaction to anesthesia. A third person told a story about a patient coming out of anesthesia who laughed so hard that everyone in the room began to laugh. I found it so odd that I could hear and comprehend everything that was being said but could not open my eyes or control my body. I later learned that hearing is often one of the first senses to return when coming out of anesthesia. After a couple of minutes, I stopped shaking and opened my eyes. Someone said, "Welcome back. You did fine." I was then asked to rate my pain on a scale of 1 to 10 with 10 being the absolute worse pain I could imagine. I said to myself, "They can't be serious. This is the worse pain I have ever experienced. Don't they know that?" But, I knew they were trying to determine how much pain medication to give me so I took the question seriously. A crazy thought then came to me of my head on a chopping block with an axe half way through my neck. I thought to myself that that would be a 10 and blurted out, "eight and a half." They gave me medication but it didn't touch the pain. Before the operation, I imagined that I would be groggy and doped up on medication while in the recovery room. I was looking forward to feeling sleepy since the feeling had become so foreign to me. I was also hoping that the operation would fix my insomnia. The reality was just the opposite. I never felt so awake. I also noticed immediately upon waking up, an intense high pitch tone in my right ear. I knew that tinnitus or ringing in the ears was a very common symptom of Chiari but I never experienced it prior to surgery.

After 30 minutes or so, the surgeon who assisted Dr. Tew approached me and introduced himself as Jonathan Sherman. I didn't have my glasses and couldn't see his face. He said that they were able to give me a really good decompression. He then went on to say that the operation took a lot longer than anticipated for three reasons. First, after they knocked me out, they realized that I was too tall to fit on the operating table. It was a special table with a head brace and it was important that my feet be placed firmly on a stand at the bottom of the table to minimize movement. He said the operation was delayed about 45 minutes until they were able to rig something up. I found it amusing because I provided the information on how tall I was at least three times to the hospital including that very morning. Next he said that my malformation was much worse once they opened me up than they had anticipated from the MRI scans. As a result, they had to wait for the lab to prepare a larger bovine graft. I remember immediately thinking to myself, "No shit, Jack. What did they think I was trying to tell them all these months?" Third, when they were closing, my brainstem misinterpreted my body temperature and I shivered so hard that I broke one of the screws on the head brace. My head fell forward and I hit my nose on one of the bars of the brace. They had to wait until they could get an x-ray machine into the room to determine if my nose was broken. The screw in the front left side of my head also tore my scalp off as my head lunged forward and they had to staple my scalp back on. Other than that, it all went well. I lay there and thought how it all made sense. The anatomical MRI for whatever reason didn't reveal the true extent of the malformation. And, also just how important it was to get a Cine MRI when a patient has a borderline herniation with serious symptomatic complaints.

Chapter 6
Early Recovery

After another 30 minutes, Dr. Tew came back to the recovery room with my family. My mother asked him what exactly was done. He chuckled and replied that they just had to make some extra room in my head because my brain was too large. He then told my family to stay as long as they liked and left. My wife had my glasses and asked me if I wanted them. I took them from her and tried to put them on but my nose was too swollen. I then noticed blood on the left side of my face and neck from where my scalp had been torn. My wife asked a nurse for assistance and they cleaned my face. The nurse explained that the plan was to place me in the intensive care unit for the first 24 hours but there were no rooms available in the ICU and I was breathing well so they were going to keep an eye on me in the recovery room until they could prepare a room for me in the neuroward. I told my family to go and get something to eat and come back later.

By 7 p.m., I was still in the recovery room. All of the other patients from the morning round of surgeries were gone by early afternoon and the last of the patients from the afternoon round of surgeries was in the process of leaving. My family returned and asked why I hadn't yet been taken to a room. Apparently, the neuroward on the fourth floor was also full. Since my family was upset, they found a room for me on the eighth floor. When we got to the eighth floor, my family saw that it was a double room and reminded the nurse that I had requested a private room. They did more checking and determined that a private room had just opened up in the neuroward. By the time I got to my room, it was 8:30 p.m. My family stayed a little longer and then left for the night. It had been a long day for us all.

About 9 p.m., Dr. Sherman stopped by to see me. He asked me to grin and show my teeth. He then asked me to raise my left leg. I did. Then, he asked me to raise my right leg. I tried but couldn't. I told him that something was holding it down. There were cuffs on my lower legs, which were slowly contracting and loosening to prevent blood clots. I told him that the cuff on my right leg must be tangled or something. He said nothing was holding it down and to try harder. I tried with all my might and managed to raise my leg a couple of inches. He then went for broke and said that he was going to help raise me up to the side of the bed and ask me to stand. I stood up and told him that I was going to faint. He immediately lowered me down on the bed and told me that he was pleased. He said that it would be much easier for me to stand up tomorrow. I told him that the pain and numbness in my face was gone but I couldn't tell about anything else. He told me to expect to experience some of my old symptoms as well as some new Chiari symptoms over the next few months. He said that they would cycle and hopefully gradually diminish in intensity with each cycle until they disappeared altogether. He also told me that I probably wouldn't be satisfied with the strength in my arms for 9 to 12 months but didn't say why. With that, he left and the night shift began.

I was permitted to have morphine every 4 hours but was encouraged to go longer without it if possible. All I had to do to get my medication was to press the call button on the side of my bed. The pain in the back of my head and neck was intense so I kept a close eye on the clock. About two in the morning, I pressed the call button for the nurse. Fifteen minutes later she hadn't come so I pressed it again. I kept pressing the button but got no response. At four in the morning, the nurse came in and asked if I needed anything. I told her that I needed morphine. She left to get it but didn't come back until an hour later. I had waited 7 hours for medication and didn't sleep at all that first night. Early in the morning, a familiar face entered my room. It was Dr. Tew's chief resident. He acted like he didn't remember me. I wanted to remind him that I was the guy whose symptoms weren't caused by Chiari but bit my tongue. He asked if I was in pain. When I told him that I was in a lot of pain, he said that he would order steroids and a muscle relaxant just for one day to get me over the hump. In addition to the pain, the anesthesia shut down both my bladder and bowels. I couldn't urinate and was becoming constipated. Every so often a nurse would come in to straight cath me and a stool softener was added to my medication regimen. My family arrived late morning. I told them that the nursing was poor and about how I kept pressing the call button all night but got no response. They stayed with me all day and would get the nurse for me when I needed morphine. I managed to eat a little food that day as well as get up and walk a few steps. During the afternoon, the nurse came in with the muscle relaxant.

It made me drowsy but not enough to fall asleep and its effect wore off in an hour or two.

Saturday night was crazy. There was a lot of activity on the ward and the nurses were very busy. I again had trouble with getting a response to the call button. When the nurse came in, I told her that I had been pressing the button quite a lot. She then asked which button and pointed out that there were two. The one I was pressing that was built into the bed rail didn't work. She took the one that did work which was on a separate cord and placed it in my lap. Thereafter, the nurse would readily respond to my call but would still take an hour or so to come back with the morphine. Instead of getting morphine every four hours, I was getting it more like every six hours. Once again, I wasn't sleepy so I simply closed my eyes and tried to rest as best I could. About three in the morning, the man across the hall began to call out loud for the nurse. He continued to call but no one responded. He then began to call out for God. He pleaded with God not to let him die without his family. His calls continued unanswered for about an hour and then finally stopped. It was awful to hear someone in such distress. I didn't understand why the nurse didn't respond to his calls sooner. It was obvious that one couldn't rest in the hospital. The objective was clearly to get well enough to go home and recover there.

Two nights had gone by and I still wasn't the least bit sleepy. I was hoping that the surgery would cure my insomnia but it seemed to only make it worse. I thought I might be having trouble sleeping because of the pain or perhaps the steroid I was given on the second day was interfering with sleep, as they are known to do. Surely, as I healed, it would improve. Sunday morning came around and I was able to make my way to the bathroom and shave and shower. I let the warm water run down the back of my head and gently rubbed the dried blood from the side of my head where my scalp was torn. My family returned by late morning and a couple of friends from work also stopped by to visit. My appetite was improving and I was able to walk down the hall to the lounge. There was something different about my legs. I could walk normal again without feeling ill. My right arm however was in pain and my left arm would hurt whenever I attempted to raise it. I was also having difficulty swallowing food. I recalled what Dr. Sherman had said about my arms and experiencing new symptoms and figured it would all work out in time.

Later that afternoon, Father Al came to visit and we prayed together to ask for healing. I thought about Psalm 18 and how my voice reached God's ears in the darkest hours of my distress. Once again, he reached down and saved me because he loved me.

Sunday night, I finally fell asleep around midnight but was awaken at 3 a.m. by an aid trying to take my vitals. I couldn't believe it and couldn't wait to get home where I would be allowed to rest.

Monday was a busy day. In the morning, I was taken to the physical therapy center to be evaluated by a therapist. He tied a strap around my waist and holding on to it behind my back we walked around the room. As we walked he gently pushed against one of my shoulders to see if it would throw me off balance. It didn't. I was also asked to ascend and descend a small staircase, which I performed with relative ease. I was being evaluated for balance because many Chiarians have balance problems from the compression to the cerebellum. Fortunately, I was spared this nuisance and would not require a walker to go home. I was also beginning to urinate but still could not void my bladder without a catheter. My nurse showed me how to catheterize myself and gave me instructions to do it once a day at home until the volume of urine collected was less than 50 milliliters. It was very difficult to do since I could barely bend my neck to see what I was doing. It was also important to perform all the steps correctly to minimize the risk of acquiring a urinary tract infection. I still however had not yet had a bowel movement.

Later that afternoon, Dr. Sholiton and her receptionist, Kathy, came to visit. Her office was only a couple of blocks away but I am sure she would have come regardless of the distance. She admitted that I was right and told me that I was only her second patient to actually have a physical problem with their brain. She told me to be sure to write a letter to the ENT she referred me to, to let him know that he missed the diagnosis.

Monday evening came and went without any sleep. By Tuesday, I was taking short walks down the hall several times throughout the day. I wanted the nurses to see my progress so I could be released to go home.

Another friend from work came to visit and I forced her to walk down the hall with me as well. Dr. Tew and his chief resident stopped by to see me later in the afternoon. I told him that it was already easier to walk and that the pain in my face was gone. He was pleased to hear that these symptoms improved. When I told him about the pain in my arms, the ringing in my ears, and the difficulty I was having swallowing my food, he indicated that I needed to give myself time to heal. He then gave me instructions on how to do isometric exercises to help speed healing of the neck muscles and told me to walk outside every day when I got home. He said to start with short distances and gradually increase the distance each day until I could walk 2 miles. Lastly, he instructed me not to hold my head in any one position for too long particularly when reading a book or working at the computer. Just to see the reaction, I then decided to compliment his chief resident and told Dr. Tew that he had helped me a great deal with pain management earlier on Saturday. They remained another couple of minutes to chat and then left.

Despite my best efforts, by early evening, there remained some concern about letting me go home. I still had not yet had a bowel movement. I couldn't void my bladder. And, I was running a low-grade fever, something to be expected after major surgery. They decided to keep me one more night and my family left about 8 p.m. At 9 p.m., the doctor on the floor came in and asked if I wanted to go home. I figured they needed the room for some poor chap who was sitting in recovery all day. Even though my family had just left, I said yes. I called home and left a message on the answering machine to turn around and come back for me when they got home.

The ride home was very unpleasant. Every bump in the road produced intense pain in my head but I was happy to be going home and was looking forward to getting some sleep in my own bed. My wife set up a small table next to the bed with water, my plastic bottle for urinating, and a bell to ring for assistance, as my voice was still very hoarse from being intubated. There were lots of flowers and plants from friends and coworkers. A professor from my alma mater even sent me a book on CD by my favorite author, James Gleick. My wife helped me change into my PJs and to get into a comfortable position in bed. My head had to be elevated with several pillows for a few weeks. Settled into my own room, my emotions began to bubble up. I never felt so helpless and dependent on others. I was still outraged by what I had gone through over the last year and a half. I was uncertain and worried about my recovery. How much strength would return? Why did my arms hurt so much? Why did the insomnia seem worse? If fatigue did not improve without sleep, how long would I be able to endure and function productively? Would the ringing in my ear subside? At that very vulnerable moment, my mother came in to see me. She ran her fingers through my hair and began to cry. She said that she wished she could trade places with me. I cried in return saying the last year had been unbearable. I kept telling the doctors I was sick but no one would help me until I was reduced to a mere shadow of myself. I thought about how dramatically medicine had changed since I was a boy. Doctors no longer knew their patients. My doctor should have known it was out of character for me to complain. But he didn't know me as a person. The health insurance situation sucked. Several doctors along the way made mention of doing an MRI but said it was expensive. I also pondered about how unlucky I was. Why couldn't I have gotten a disease that doctors understood? Nevertheless, it was time to put all the bad that had happened to me behind me, and focus on recovery. I had promised Dr. Tew that if he did his job, I would do mine.

Pain was the main problem during my first week at home. Lowering my head to any degree produced a splitting headache. The Roxicet (morphine/Tylenol combination tablets) didn't seem to help very much. I called Dr. Tew's nurse, Nancy, to ask if I could take some additional Tylenol for the pain but Dr. Tew said no. I only had enough Roxicet for a week. After that, I could only use OTC analgesics. After the Roxicet ran out, I expected the worse but to my surprise, I got greater relief when I took Advil. I continued to catheterize myself every day until the volume collected was below 50 milliliters, which took about one week. Going out for my daily walk was a major event. My wife had to assist me in putting my pants and shoes on. Holding on to my wife's hand for support, I only walked about 100 yards the first day. Other than using the bathroom, I spent the rest of the day in bed. I watched TV, listened to music, and did crossword puzzles. As much as I wanted to nap, I couldn't. Sleeping remained a problem. I was only sleeping a few hours every other night. I also expressed concern to Nancy that week about continuing to run a fever and the insomnia. My temperature would be normal in the morning but climb throughout the day until it reached about 101 in the evening. Nancy told me not to worry about my fever unless it went above 101. As far as insomnia went Nancy said that others found it easier to sleep after they got their stitches out.

By doing my neck isometric exercises on a regular basis, the pain subsided after a couple of weeks. I increased the distance I was walking each day as instructed but the low-grade fever continued and I was cold all the time. After 10 days, I returned to Dr. Tew's office to get my stitches out. He looked at the back of my head after Nancy removed the stitches and said that it appeared to be healing well. I told him how I was cold all the time. He said that it must be a metabolism thing and suggested that I walk outside twice a day to raise my metabolism. By four weeks, I was walking 2 miles a day. Clearly, the strength had returned to my legs. Overall, however, I began to feel that I traded one set of symptoms for another. My arms, which were weak before, were now also in pain. The pain in my face was gone but I was having trouble with food getting stuck in my throat. I could walk without feeling ill but my feet and hands were falling asleep all the time for no reason. I was still having trouble sleeping and with frequent urination. Despite my new problems, I was making forward progress. After 4 weeks, I was no longer in bed and would sometimes go to the mall for my walks. I began to drive again at 5 weeks. One day while taking a walk, I decided to see if I could jog so I did so intermittently over the course of about 2 miles. The result was a disaster. I was unable to sleep that night and became sick for a day. From that experience, I decided that when it was time to resume exercise, I would start by riding my bike. At 6 weeks, I returned to Dr. Tew's office for my final follow up exam. He was pleased that I was able to walk so well, after all, the goal of the surgery was simply to prevent further decline. I told him that I want to go off disability and return to work part time the following week, to which he replied, "Now, that's real progress."

Chapter 7
Back to Work

Seven weeks postop, I returned to work, working half days. My arms were too weak to carry my briefcase so I acquired a fold out cart to roll it from my car to the office. On days where I slept the night before, I felt pretty good, particularly in the morning. Even on good days, I was fatigued by early afternoon. In January, I selected a new primary care physician in my health insurance plan. I wanted to establish a relationship with her and inquire if she could prescribe some physical therapy for my arms. When I met with the rehabilitation doctor, he told me that he had never worked with a Chiarian before but would approach it like an upper spinal cord injury. He also obtained all my records from the hospital and provided me with a copy as well. In addition to the 2-page surgeon's report, there were another 50 or so pages covering the details during my surgery as well as various radiology reports. I was surprised to read that the x-ray that was taken during my surgery to see if I had broken my nose was really ordered for an incorrect needle count. Further, the radiologist's report indicated that the results were inconclusive and recommended a follow up x-ray. About this same time, I felt a sticky fluid with my hand oozing from the incision in the back of my head one day at work. I asked my secretary to take a look at it and she indicated that it appeared to be infected. I shook my head in disbelief and said, "My God, they left a frigging needle in my head." I immediately called Nancy, who instructed me to come in right away. She inspected the incision and said that it appeared to be superficial and being caused by either a stitch that was missed or some dissolvable stitch that worked its way to the surface. She removed the stitch and cleaned the area. I then went downstairs for another x-ray, which was brought immediately to Dr. Tew. He read the film and assured me that no needle was in my head. He apologized for the broken nose/incorrect needle count discrepancy and indicated that he was going to have a word with his subordinates. From this, I derived that he must have left the operating room towards the end of the procedure and instructed his surgeons to close without him. What really happened to me during closing remains a mystery. Doctor friends at work who I told the story to, were of the opinion that mere shaking couldn't break a head brace, that something more violent, like a seizure, was required.

Over the next couple of months, I went to physical therapy and gradually increased my time at work until I was once again working full-time. The physical therapy was unsuccessful. When I used my arms to any extent, the pain would increase and my arms would also get the sensation of pins and needles. The therapist was interested in determining if the reaction was related to my heart rate by having me peddle an ergometer while measuring my pulse. No relationship could be established. After 10 sessions or so, the therapist told me that they couldn't do anything more for me. Oddly, she then suggested that I seek an alternative healing technique known as cranial sacral therapy. I called a practitioner but expressed some skepticism. He suggested I read a particular book first before deciding to try it. I bought the book, read about half of it, and decided the technique did not have a valid scientific basis. Strangely enough, I decided one day to get a full body massage, which was available at the fitness center at work. I was hoping that it might help to relax me and improve my sleeping. At the beginning of the massage, the masseuse mentioned that she was also trained in cranial sacral therapy and offered to perform it on me after the massage. The technique basically involves running the hands over the surface of the body as a way of detecting pulses in the body's clear fluid indicative of flow restrictions. If a restriction is detected then a gentle manipulation is performed to correct the restriction or imbalance, which in turn relieves the symptoms. In my case, the masseuse found no restrictions. She indicated that my system was in good harmony.

While I was walking much better, the insomnia and fatigue was extremely difficult to battle. I was also urinating frequently and slowly losing weight. And, while I was cold all the time, the bottoms of my feet always felt like they were standing on a hot beach. I continued to work with Dr. Sholiton on my sleep problem. She asked if she could treat me for anxiety and I agreed to add Ativan and BuSpar to my regimen, medications that I had taken before. The combination of Remeron, Klonopin, BuSpar and Ativan just didn't work for me. It made me sick. We talked further. I expressed my opinion that I wasn't suffering from anxiety. Not being able to fall asleep might be a sign of anxiety but I told her I knew what anxiety felt like and I just wasn't there. I suggested that we treat the insomnia more directly and asked if I could try Ambien, a sedative approved specifically for insomnia. I had some immediate success with Ambien so I quickly dropped the Ativan because I saw it as redundant to both Klonopin and Ambien. The Ambien worked for a month or so and then I adapted to it and returned to the land of insomnia.

I also returned to my new primary care physician because I wasn't certain where to go for follow up care. The surgeon had done his job and from what I read on-line, most neurologists knew little about Chiari. Because I was urinating frequently and losing weight, I thought that she should test me for diabetes. I told her that I had gone to the pharmacy and purchased a self-diagnostic test for urinary tract infections (UTI) and that the results were negative. She tested for UTI again in her office and then exclaimed that my case was very complicated and suggested I seek follow-up care with Dr. Tew. It is important to understand that the medications I was taken were capable of causing frequent urination but frequent urination started long before I was on any medications and it did not occur if I managed to get some sleep.

Chapter 8
Re-evaluation

By now it was June 2000. I called Nancy and explained that I wasn't doing well and that Dr. Tew might want to consider re-evaluating me. She discussed it with Dr. Tew and called me back. Dr. Tew thought it best to refer me to his neurology colleagues across the hall for a complete neurological work up. When I called to make an appointment, they were booked solid for 2 months. I pleaded to get in sooner and was told that a Polish neurologist who was there doing a rotation would be able to see me a couple of weeks sooner if that was OK with me. I agreed to see him. I found the Polish neurologist to be both thorough and generous with his time. He ordered an anatomical and Cine MRI, an EMG and nerve conduction tests for my arms, and a large amount of blood tests. He wanted to rule out complications from the surgery such as the formation of scar tissue, carpal tunnel and nerve damage to my arm, diabetes and host of neurological diseases capable of producing similar symptoms. Importantly, he also suggested that I have a sleep study done to rule out central sleep apnea and other sleep disorders and referred me to Dr. Bruce Corser.

When I went to schedule the MRI scans, I found I could not return to the university facility, which did my initial scans because they were no longer covered by my health insurance plan. I had to go to a different hospital and radiologist and make arrangements to have my old MRI scans pulled from the archives and sent to the new hospital for comparison. This proved to be a daunting task. I had a personal copy of my original anatomical MRI scans but I did not have any Cine MRI scans and the university was not able to locate them. At any rate, the follow up MRI results looked good. There were no new findings in the anatomical MRI and it showed that considerable space was created by the decompression. Likewise, the Cine MRI confirmed that nothing was blocking the flow of CSF. The EMG and nerve conduction tests were also negative as were all the blood tests. All of this was good news but served to increase the importance of the sleep study results.

When I went to make an appointment at the sleep clinic, I again ran into scheduling problems. Dr. Corser was in high demand and it would take several weeks to get in to see him so I agreed to see his associate, Dr. Armatage. Before coming in to see Dr. Armatage, I was asked to complete a 2-week sleep diary. It showed that I was only sleeping about 3 or 4 hours every other night and that my numerous attempts to nap during the day were futile. Dr. Armatage reviewed the diary and asked me several questions pertaining to my brain defect and the surgery performed. He then discussed why doing a sleep study in the presence of medication would only produce results of limited value. We worked out a schedule for weaning me off the medications over a two-month period before coming back for the study. While attempting to wean off the medications, the insomnia became unbearable. I had grown dependent on the medications for what little sleep I was getting. I called the clinic back, explained my situation and talked my way into seeing Dr. Corser. Dr. Corser felt that I had one of the worse cases of insomnia he had ever seen and wanted to proceed with the sleep study with the medications on board. He then mentioned that he had heard that a drug for bipolar disorder, Zyprexa, was good for refractory insomnia. A couple of days later, I had an appointment with Dr. Sholiton and covered the next steps I would be taking with Dr. Corser and the sleep study. When I mentioned the Zyprexa, she asked if I would like to try it. I indicated that I would so she prescribed it. Before taking it however I discussed it with Dr. Corser. He suggested that I start with the smallest dose of 2.5 mg. I took 2.5 mg but it failed to put me to sleep. I called Dr. Corser the following day and he suggested I take 5 mg. This dose also failed. The third night he suggested 7.5 mg, which finally put me to sleep. The sleep study, which was now scheduled for later that week, would now be further compromised by the inclusion of Zyprexa to the psychotropic cocktail I was taking.

While all this was going on, I was arranging for a second opinion by Dr. Thomas Milhorat in New York, who is recognized as the foremost authority in Chiari in the country. I had discussed this on several occasions with Dr. Sholiton. Her opinion was that I would never obtain peace of mind until I consulted with the best. When I called his office, his nurse spent a very generous amount of time with me on the phone to understand my situation. She stated that Dr. Milhorat preferred patients who have already been decompressed to first see his associate, Dr. Roger Kula. Dr. Kula was a neurologist with a keen interest in Chiari. I called Dr. Kula's office and arranged for a visit in October. They asked me to bring all my test results in along with the

MRI scans and the sleep study I was about to under go. In addition, they were to send me some fibromyalgia and chronic fatigue syndrome symptom questionnaires to complete and return at the time of my visit.

Time for the sleep study arrived. I checked into the clinic and was quite impressed with the room accommodations. It was equivalent to a nice hotel room. The technician applied a number of sensing devices to my body to monitor my breathing, heart rate, muscle and brain wave activity. I took all of my medications and went to bed. The Zyprexa kicked in quickly and I went to sleep for approximately six hours, making it a legitimate study. During my follow up visit with Dr. Corser, he noted the absence of central sleep apnea. Central sleep apnea is a condition where the individual stops breathing during sleep and awakens for a brief moment. These interruptions can occur frequently during the night leading to un-restful sleep. Central sleep apnea has its origin in the brain stem and it is observed frequently in Chiari due to compression of the brain stem. This was the primary condition Dr. Corser was checking for in me. Since drugs usually don't affect this condition, he wasn't worried about them interfering with the study. However, he did note that I kicked my legs repeatedly throughout the night, a sign that I might be suffering from restless legs syndrome (RLS). In addition, I was found to get no phase 3 or 4 sleep and an insufficient amount of REM (rapid eye movement) sleep. Net, when I did sleep, I was too busy kicking around and never entered the deeper levels of sleep important for achieving a good restful sleep. I told Dr. Corser that I was somewhat skeptical of the RLS diagnosis because I noticed my legs began to twitch only after I had begun taking Zyprexa. Drugs known to increase the level of dopamine in the brain can successfully treat RLS. Dr. Corser suggested I try one but I declined because I suspected Zyprexa as the cause and I was already taking too many psychotropic medications.

Next, it was back to the local neurologist for a follow up visit. When I returned, the Polish neurologist was off doing a rotation somewhere on neuroimaging so one of the other neurologists in the group saw me. He went over my entire test results, which were negative. He agreed that sleep would help but couldn't rule out the possibility that I was just dealing with damage caused by Chiari in the first place. He offered to write me a prescription for sleep. When I told him all the medications that I had tried, he wrote out a prescription for Thorazine and instructed me to use it only when I was at my wits end. I didn't like the idea of graduating to a new level of more powerful drugs so I never got the prescription filled.

A month later, I went to New York to see Dr. Kula. One of his residents spent a couple of hours with me collecting a medical history and the details of my Chiari symptoms and surgery. When Dr. Kula entered, I ran down the list of my current complaints - insomnia, fatigue, general coldness, frequent urination, weight loss and burning feet. He gave me a complete neurological exam, which I passed. He reviewed the completed questionnaires and all of my test results paying close attention to the MRI scans and sleep study results. He also had his resident review my medical history. In all, Dr. Kula spent an additional hour and a half examining me and reviewing all of the information. In the end, Dr. Kula was of the opinion that my decompression was well done and that I was obviously not depressed. He felt that, if anything, I was suffering from RLS and if my sleep problem could be treated with less medication, my symptoms would resolve. He was also somewhat concerned that an MRI of the spine was never performed to rule out Syringomyelia. He wrote out an order for three things, 1) decrease current medications, 2) go on Mirapex, a dopamine agonist for RLS, and 3) get an MRI of the spine.

When I returned home, I shared Dr. Kula's orders with Dr. Corser. I decided this time to try Mirapex. Dr. Corser ordered the spinal MRI and we discussed a strategy for reducing my other medications. The Mirapex did nothing to either help me fall asleep easier or to sleep better on the rare occasions when I fell asleep. The MRI showed no evidence of a syrinx. As far as reducing my other medications, I failed to make any progress. I had grown dependent on them and every time I tried to cut back on one of them, the rebound insomnia was more than I could handle. At this point, I did all I could to find additional answers. I was checked head to toe by an excellent local neurology group. I was re-checked by Dr. Kula in New York. I had a sleep study while newly dosed up on Zyprexa. The Zyprexa, which helped me get a little sleep for a couple of months, lost its effect.

By December of 2000, I decided to join the fitness center at work with the objective of strengthening my arms. The constant pain that I was experiencing in my right arm had resolved but my left arm still hurt when I tried to raise it while bent at the elbow. I consulted with one of the fitness staff on what apparatus would be best and selected the lateral press. I started with a low weight and increased the load gradually. As time went on, I added other complementary exercises until I was using 6 different machines routinely for my arms, back and chest. Over the course of a few months, I increased my strength significantly and the pain in my left arm subsided completely. The bigger problem that I faced however was the feeling of illness that I would get upon aerobic exertion. Aerobic exertion also seemed to significantly worsen my insomnia, which I found very puzzling. One would think that aerobic exercise would improve sleep. I decided that I should challenge my body but go gentle at first. I figured if I keep sending signals to my brain, it would eventually interpret them correctly. Jogging was too much so I decided to walk on a treadmill at the fitness center during the workweek and ride my bike on the weekends. I worked at it for several months. Riding the bike was more difficult but over the months I managed to extend my weekend ride to 25 miles. There seemed to be a magic line. If I stayed under a certain exertion level, I wouldn't get ill, but I had a great deal of difficultly judging if I had crossed the line. Also, the illness did not come on until a couple of hours following the exertion. The 25-mile bike ride always made me ill but I persisted.

Chapter 9
Summer Camp

The first 6 months of 2001 were in many respects as difficult as the months preceding the decompression. I wasn't sleeping and was constantly fatigued. Despite this, I persisted to exercise and work full time. However, after work, I had no life at all. I was no longer participating in the local American Chemical Society meetings. I stopped attending weekly scout meetings and most scout activities. We committed to no social events. I found, that in order to have any chance at all of falling asleep; I had to take my medications on an empty stomach so I didn't eat any food after 6 p.m. As a result, my routine was very rigid. Work became an objective of survival. I wasn't able to concentrate and had no tolerance for details. I was fortunate to have great subordinates reporting to me who recognized my situation and covered for me on all fronts. While I was making some progress with my arms, I continued to feel ill all of the time. After many months of insomnia, depression began to creep back into my life. By the middle of the year, I would find myself thinking how desirable it would be to drive off the road on my way home from work. The quality of my life wasn't improving. As much as I didn't want to believe it, I thought how the surgeons told me that they could only promise to stop my deterioration with decompression. My hope for improvement was fading. Perhaps they were right. Some symptoms like tingling, burning and facial pain were gone but on the whole I hadn't managed to improve. I couldn't see going on without sleep. I didn't like the fact that I was taking so many medications, which also had their numerous side effects. With each drive home the temptation of driving off the road strengthened. I also found myself preoccupied with events 25 to 35 years ago for which I had regrets. There were people who I may have offended as a teenager who were now haunting me. I realized my thoughts didn't make sense but I couldn't control them.

I called Dr. Sholiton on a Friday afternoon in June and told her that I was in trouble. She asked me if I wanted her to admit me somewhere where I could be safe. I told her that I did and she asked me to come to her office. I had my wife drive me to her office. I told Dr. Sholiton that I needed something more than a psychiatric ward. I also needed a hospital with a sleep center. My problem wasn't that I couldn't sleep because I was depressed. My problem was that I was depressed because I couldn't sleep. Dr. Sholiton found a local hospital with both a psychiatric ward and a sleep center and I was admitted that afternoon.

I couldn't believe that I had sunk so low after all I went through to try to get better. I was now in a psych ward without a belt on suicide watch. I did not sleep the first night. The next day I met with one of the psychiatrists on duty. I told her about my history with Chiari and how my inability to sleep had forced me into a state of depression. We discussed the medications that I was taking. Obviously they weren't doing me any good so she ordered me off the meds with the exception of Klonopin, which cannot be abruptly withdrawn. A practitioner of some sort also visited me that day and taught me a couple of relaxation techniques. One technique involved rubbing my skin gently with a soft brush. The other came with an instruction tape and involved a systematic way of tightening and relaxing muscles from head to toe. I thought both were a bunch of nonsense because I wasn't tense. I could get into a very relaxing state of mind. I just couldn't fall asleep. That weekend I engaged in all the activities in the ward and laid awake in bed on both Saturday and Sunday evenings. Early on Monday morning around 4 am, I had a panic attack. I had had no sleep for 3 nights on top of being abruptly withdrawn from Zyprexa, Remeron and Ambien. A nurse came to my room and in a very stern voice told me that I was the only one who could help me. She suggested I turn on my relaxation tape and go through the muscle exercises. I did and it worked to my surprise.

Mornings in the ward were interesting. It seemed like the place had no energy anywhere to get started. Most of the patients had to be pried from their beds for breakfast. The staff also seemed late in getting started. Most mornings began with a light exercise session after breakfast. While in the session, patients would be called out to meet with their doctor or other caregivers. On, Monday morning I was called out of the exercise session to meet with the regular psychiatrist that I was referred to by Dr. Sholiton. He was very nice but we butted heads to some degree when I requested a consult with a sleep specialist. He indicated that there wasn't a sleep specialist at the facility who participated in my insurance plan. I told him that I didn't care and would pay for the consult out of pocket if necessary. I told him that I had been admitted to the hospital with the distinct understanding that I would receive care from both a psychiatrist and a sleep specialist. At that point, he indicated that he would see what he could arrange. After meeting with my doctor, I returned to the group. After exercise, there was a group therapy session followed by lunch. In the afternoons, there was another group therapy session and either an arts and craft class or ceramics. It was very much like summer camp and an ideal stress free environment for an insomniac.

In the afternoon, I was pulled out of a session to see the sleep specialist. I told him about my problem and brain surgery and how difficult it was for me to sleep. I told him how Dr. Corser did a sleep study on me and diagnosed me with idiopathic insomnia and suggested that I try Zyprexa because he had heard that it was effective for refractory insomnia. The sleep specialist basically told me not to worry. That he had seen many patients similar to me. He indicated that he would contact Dr. Corser's office and obtain a copy of my sleep study to review. He agreed that I was taking too many medications and suggested that I obtain a copy of a book entitled, "Desperately Seeking Snoozin" by John Wiedman. The book talks about curing insomnia by first going off all medications and then adhering to a very short sleep cycle, which is gradually lengthened as the mind re-learns how to sleep.

On Tuesday, I was extracted from the morning exercise period to meet with both my psychiatrist and sleep specialist. The psychiatrist indicated that he wanted to put me on Anafranil, a tricyclic antidepressant, effective for obsessive-compulsive disorder. He believed that I was obsessing about not sleeping and the drug would help. In addition, the drug was known to be sedating. I also pleaded with him to allow me to continue taking Zyprexa because I felt it was the most effective drug that I had taken for my insomnia. He reluctantly agreed and stated that he wasn't aware of any adverse interaction Zyprexa might have with Anafranil. While the sleep specialist preferred that I come off drugs completely and try the desperately seeking snoozing method, he readily agreed that I needed something for my obsession and that after things got under control, I could go off the medications and try the more natural approach toward curing my insomnia. I expressed concern about the potential side effects of Anafranil since the tricyclic antidepressants in general were known to have more side effects than the modern antidepressants. The psychiatrist said that in his experience those who needed the drug most tolerated it the best. I was to start with a 50mg dose and work up to 100mg before leaving the hospital.

That evening, I took a 50mg dose of Anafranil. It didn't make me drowsy and I did not sleep. However, in the morning, I had difficulty urinating. It was as if I had no need to urinate. This was good since I had been suffering so long from frequent urination. More important and to my surprise, I was no longer cold. Later that morning, I called my wife and asked her to bring me short pants. On Wednesday evening, the dose was increased to 75mg. Again, I did not sleep but my appetite returned the following day. On Thursday evening, I took 100mg and again was found to tolerate the dose well.

Life in the psych ward was interesting to say the least, particularly some of the other people that I met. Each had their own story and their suffering was truly deep. I would come to respect all of them. I was particularly drawn to Cindy. We both had similar surgical scars on the back of our heads. She had an aneurysm but her outcome was the same as mine. She could not sleep. We would see one another off and on throughout the middle of the night as we walked the halls, watched television and worked on jigsaw puzzles. Being abruptly pulled off most of the sedating drugs I was taking resulted in rebound insomnia on top of my insomnia problem. For the week that I was in the ward, I had only slept a total of 5 hours. I was the king of insomnia and spent most of the nights after the initial weekend working on several jigsaw puzzles. As I worked on the puzzles throughout the night various other ward residents would wonder in and join me. As a result, I got to know many of them. Some were there for major depression episodes, some were in alcohol/drug detox, and some were clearly psychotic. I didn't get to know the patients suffering from psychosis. They would usually sit in an area or corner away from the rest of us, uttering to themselves. One was a beautiful young lady who sat in the corner listening to the radio. She would change the channels frequently and had to be told often to turn the volume down. One night, she hastily walked into the living room and began taking large numbers of books from shelves adjacent to the television. The night nurse intervened. She told the nurse that she had to read the books. The nurse tried to reason with her by telling her that she didn't have enough time to read so many books. They ended up compromising, allowing her to take about 6 to 8 books back to her room. The next morning, everyone on the ward could hear her screams in the halls as several staff tried to subdue her. I thought about how tragic her situation was, to be so young and beautiful and have no control of her emotions whatsoever. I assumed she wasn't married because no one ever came to visit her. I thought about how she should be married with children and a career and a happy life in general. Instead she was strapped down in her room, sedated, and without any peace of mind.

There was another young beautiful married young lady in her early twenties suffering from depression and insomnia. When she first came in she tended to sit by herself, appearing to be more afraid than psychotic. The jigsaw puzzle group coaxed her over to the table. She had a knack for puzzles and quickly made friends in the group. Although looks can be extremely deceiving amongst patients in a psych ward, she seemed to be the type who needed some attention and tasks to do in order to get her mind off her troubles. She seemed to be greatly improved after a couple of days.

An older lady, about my age, also joined our puzzle team. She was hooked on Oxycontin and up half the night in withdrawal. Other than me, she was probably the most consistent member of the third shift puzzle team.

Cindy wasn't much for puzzles, as she couldn't concentrate, a common problem with patients on the ward. I mentioned one evening while watching television with her across from the ping-pong room that

I used to play table tennis at the college level. She immediately perked up. Who would have guessed that we had a common interest in ping-pong? We checked out some paddles and a ball and began to play. She was extremely competitive and it wasn't long before we drew a small crowd. Others showed an interest in playing so we fortified the poorly mounted net with masking tape and started a tournament.

As I got to know the other patients and learned that I would not fall apart without sleep, my situation began to seem trivial and my own thoughts of suicide turned into thoughts of determination to live life again even if my body was in a constant state of fatigue.

On Friday, one week after being admitted, the doctor recommended that I be discharged. I was pleased with how the side effects of Anafranil counteracted the problems I was having with frequent urination, coldness and weight loss. I wasn't impressed with the sleep specialist consultation but learned that I could tolerate a much greater level of insomnia than I had believed. I also understood that the benefit of Anafranil on depression and my obsessiveness with insomnia would not kick in for several weeks. Anafranil is usually sedating, perhaps due to its antihistamine properties, but it did not make me sleepy in the least. The rebound insomnia from discontinuing Ambien and Remeron may have overwhelmed Anafranil's sedating effects. Overall, being in the hospital was a good thing. It was a stress free environment, which made discontinuing medications that weren't really helping me go easier. I also realized that I wouldn't break if I didn't sleep for 4 or 5 days.

I was discharged from the hospital on Friday and returned to work on Monday. Things were difficult for the next 6 weeks or so. I got very little sleep and didn't feel well. However, by fall, the Anafranil began to kick in. The depression lifted and I began to sleep 4 or 5 hours a night on a regular basis. The change in medication made all the difference in the world. I also found that my tolerance for exercise was greatly increased so I began to jog again every other day and lifting weights on the days that I did not jog. However, after a few months, my sleeping pattern began to change again. I was still sleeping 4 to 5 hours a day on average but the amount of sleep I got on any one day was highly variable. Some days I did not sleep at all. Also, I never stayed asleep for more than 2 hours. I would wake up in the middle of the night with very vivid dreams. To clear my head, I began to go downstairs to the kitchen in the middle of the night and make breakfast. By the time I would finish breakfast my mind was clear again and I would go back to sleep for another hour or so. This pattern continued for many months.

Chapter 10
Restored Confidence

By late summer of 2002, I had been doing so well for so long that I decided that I should try to cut back on my medications. While the medications provided great benefit, they also imparted many unwanted side effects such as orthostatic hypotension, dry mouth, constipation, and farting. I was also convinced that the medications were the cause of my vivid dreams and my strange sleeping pattern. I also found myself falling asleep at work in the morning while going through my email and sometimes, to a much lesser frequency, even nodding off during meetings. And, on top of all this, I gained over 30 pounds.

I was taking 100 mg of Anafranil, 7.5 mg of Zyprexa, and 1.5 mg of Klonopin. Of the three, Klonopin is the most addictive. I informed Dr. Sholiton that I was going to wean off Klonopin first as it was the most addictive and it would be easier to do with Anafranil and Zyprexa on board to control anxiety. I decreased the Klonopin at a rate of 0.25 mg per week. With each decrease, it caused some rebound insomnia but it was manageable. I went off the last quarter milligram in early September while on vacation with my son at my father's home in the east. When I first went off the drug the biggest problem I had was profuse sweating particularly with light activity such as walking. I took my son by train up to Manhattan for a day visit on 9/10/02, one day before the first anniversary of the 9/11 tragedy. It was a beautiful clear day with temperatures in the low seventies and a mild breeze. I was dressed to stay cool with a tee shirt and shorts. Nevertheless, by mid-day my shirt was completely soaked. On our way to the Statue of Liberty, we passed ground zero. I was overwhelmed with the shear size of the pit but more so by all the flowers, notes, and pictures attached to the cyclone fence around the site. There were people standing by the fence reading the notes and weeping. I could not bring myself to do the same. I said a silent prayer and continued on.

After a week of being Klonopin free, my insomnia worsened significantly so I went back to taking 0.5 mg and congratulated myself for shedding a full milligram.

In the spring of 2003, I was scheduled for my routine annual physical exam. In preparation for the physical, I lost 20 pounds because I knew my cholesterol might be an issue and I didn't want to add another medication (in this case, a statin) to my regimen. Losing the 20 pounds with the appetite-enhancing medications on board was particularly challenging but somehow I managed. All the results of my physical were fine except cholesterol, which came in at 222, down 20 points from the previous year. I told my doctor that I was reluctant to start cholesterol-lowering medications with all the other drugs I was already taking. I told him that I wanted to wean off the psychotropic medications and lose more weight over the next year and then see where my cholesterol was.

Shortly after the physical, with a great deal of determination, I decided to wean off the medications. This time the appetite-enhancing drugs, Zyprexa and

Anafranil, were my targets. I made the mistake of being way too aggressive tapering off the two drugs in only 3 weeks. I anticipated rebound insomnia but got more than I bargained for. After being awake for an amazing 8 straight days, I caved in and went back on both medications to the original 7.5 mg dose of Zyprexa and a somewhat slightly lower dose of 75 mg of Anafranil. I was very disappointed in myself for attempting such a rapid weaning schedule. I was also uncertain if my underlying insomnia condition had improved any. I decided to wait again before trying to wean off the medications again. Vacation was approaching and I wanted to enjoy it without any withdrawal symptoms.

Sleep improved upon returning to the medications but it was still far from ideal. I continued to sleep 4 hours a day on average. Any attempts at taking a nap continued to be futile. At least once or twice a week I would not get any sleep at all. On these nights, the medication seemed to stimulate me. Other symptoms persisted. Swallowing was sometimes sluggish and always produced loud pops in my ears. My left arm would hurt from time to time. Running 3 miles was my limit despite months of training. While being able to run 3 miles a day doesn't sound like a problem, it was for me since in the past I was capable of easily running 10 miles a day after a few months of training. I remembered Dr. Tew telling me before he was convinced that my Chiari malformation was causing my symptoms that there was nothing wrong with me that a couple of weeks of good natural sleep wouldn't fix. I believed him to be correct but I didn't know how to get natural sleep. Sleep was indeed the key. All of my deficits and limitations could be explained by sleep depravation. I also wanted to believe that if I could sleep with drugs, I could probably sleep without them if I could find a weaning schedule that would work.

At the end of 2003, I made another attempt to wean off the medications at slower rate. I chose to come off Anafranil first. The withdrawal symptoms did not become severe until I went completely off the drug. At that point, a weird form of dizziness set in. Every time I moved my eyes I could heard a whooshing sound in my head. I managed to maintain my running but found it necessary to focus my eyes on a single point during my runs in order to stay on the treadmill. My insomnia intensified but I managed. After a month, the dizziness dissipated and my sleep went back pretty much to what it had been while taking the drug. My appetite markedly decreased and I lost a few pounds. I also noticed that I didn't sweat as much on my runs.

With renewed confidence, I began to wean off the Zyprexa. Other than rebound insomnia, I didn't experience any withdrawal effects.

I managed to remain drug free for several weeks but I was hardly getting any sleep. I was convinced that I had gotten past the period of rebound insomnia. I was not depressed or anxious. I was not obsessed with getting sleep as many suggested. I simply did not experience sleepiness. Exercise would make me tired but not sleepy. When I did sleep a couple of hours, I felt refreshed and my thinking was clear.

I liked being free of medication but the insomnia was chipping away at my psyche. Even if one doesn't get sleepy, one still needs to escape "awakeness" psychologically.

By now, my psychiatrist, Dr. Sholiton, who had seen me through several years of depression and anxiety and brain surgery, had proclaimed me cured. She had been critical in my treatment and convinced me that many Chiarians need such support if they are ever to become well again. I would miss my sessions with her but not to the point where I wanted to remain her patient. From now on, I was a pure and simple insomnia patient.

I returned to my primary care physician who was also a sleep specialist and discussed the matter with him. We discussed several options for treating my insomnia starting with reviewing past medications tried. Trazadone caused so much nasal congestion that I couldn't breathe through my nose, a personal requirement for sleep. The benzodiazepines, like Ativan, Restoril and Klonopin were out – too addicting. Besides, they had only been effective for a few days at best. Melatonin didn't work. SSRIs increased my insomnia. I had grown resistant to Remeron. GHB (gamma hydroxybutyric acid), the so-called "date rape drug", was offered but I was concerned about its poor safety profile. OTC antihistamines didn't work. Ambien caused the jitters and wasn't very effective. After going through these drugs and few others, we decided to try low doses of Zyprexa and Elavil. This time, I would limit Zyprexa to 2.5 mg and Elavil to 25 mg. At low doses, they helped enough to keep me going with manageable side effects.

Through 2004 and the first half of 2005, I maintained my jogging program. When I tried to increase my running by running either longer distances or running faster, my insomnia would intensify. It was as though the increased running would excite my nervous system. Instead of making me sleepy, it would make me more vigilant. However, slowly and steadily, I increased the amount I could run on a daily basis.

By July of 2005, I was running nearly 40 miles a week. I had gotten to the point where I could run 7 or 8 miles and still manage to get some sleep the same night.

For the first couple of years after surgery, I thought I would never run again. Every time I saw a jogger I felt as though I had been robbed; that something had been stolen from me. It simply wasn't true. Neurological damage takes time to resolve. It also requires challenge. One must challenge the nervous system, challenge it repeatedly. By whatever means, whether by direct healing/re-growth or by re-wiring, the nervous system responds to challenge over time. I needed more time and with that time came healing and the ability to run once again.

Chapter 11
The Marathon

Mid 2005 was a turning point. Up until that time, when I ran more than 3 miles, I could not sleep at all. But this began to change. I found that I could run longer distances and manage to get some sleep the same night. This was extremely exciting and the idea of training up for a marathon began to seriously form in my mind.

It had been a long time since I had run such a distance. I was 31 the last time I ran a marathon. I completed that race in 3 hours and 30 minutes. I was now 22 years older. Could I train up without injuring myself? Would my insomnia, which still persisted to a significant degree, limit how far I could ultimately run? I figured it would be worthwhile to find out and devised a training schedule that would gradually increase my weekly distance from 25 to 30 miles a week to 40 to 45 miles a week. The schedule included one or two days of rest a week as well as strength training and stretching exercises. Distance would be increased 10% per week to minimize the possibility of injury.

I set my sights on the Air Force Marathon in Dayton, Ohio, on September 17th. The advantage to running the Air Force Marathon was that participants had a full 8 hours to complete the course. In previous marathons, I averaged 8-minute miles. Being significantly older, I set a target pace of 9 to 10 minute miles for this race which would place my finish time somewhere between 4 and 4½ hours.

In the midst of deciding to run a marathon, I decided to up the commitment. I contacted Rick Labuda at the Chiari & Syringomyelia Patient Education Foundation, also known as Conquer Chiari. I knew that Rick was looking for projects to raise money for research. I told him about the marathon and that I would be willing to collect pledges. He was very enthusiastic about the idea and featured it in the August Chiari and Syringomyelia News. I also contacted a local television station, WB64, and told them about Chiari, my struggle with it and my effort to run a marathon to raise money for research. My overture was received with great interest in part because the reporter's mother had Chiari. The reporter and her video team interviewed me, my doctor and a local mother whose child had Chiari and aired the story on the 10 o'clock news.

Following the schedule to train up for the marathon got off to a good start. I continued to be able to get some sleep as I increased my running. I found it so encouraging that I began to see an opportunity once again to possibly wean off medications. I really hated the idea of taking psychotropic medications the rest of my life. After all, my depression had been gone for several years and I was never schizophrenic. The only reason I was taking an antidepressant and antipsychotic was for their sedating properties. Although I was taking very low doses of each, I did experience unwanted side effects such as dry mouth, excessive sweating, increased appetite, and urinary hesitancy. In addition, the Zyprexa seemed to compress the range of my emotions. I never felt really happy and I never felt really sad. I couldn't get excited or angry.

I decided to try to wean off the medications again. They just didn't seem to be as necessary as they had been. Also, being off the medications, I could probably slim down a few more pounds which would be beneficial for the marathon. I also figured that I was probably producing more endorphins with my increased exercise which might help ease any withdrawal effects, particularly any transient anxiety.

I was already taking the lowest doses of both Elavil and Zyprexa, 25 mg and 2.5 mg respectively. I decided to discontinue taking Elavil first. Unlike previous attempts, I found it relatively easy. It did not increase my insomnia. I felt a little down and anxious on the second day without any Elavil but it passed within 24 hours.

After being off Elavil for a week, it was time to discontinue taking Zyprexa. Going off Zyprexa in the past had been much more difficult. By the second day, I would feel very sick and the rebound insomnia was intolerable. This time, it was different or so it seemed at first. The first few nights without Zyprexa, I found it hard to fall asleep due to minor aches and pains from all the running. However, I found that by taking a couple of Tylenols, I was readily able to fall asleep and sleep long enough (4 hours) to feel reasonably refreshed. The third and fourth days without Zyprexa were difficult. I didn't feel well at all and I was experiencing significant gastrointestinal distress. I was having about 5 bowel movements a day including a movement at 4 a.m. In addition to hitting me in the middle of the night, it would even hit me while running. On the fourth day in the middle of my run at 5 miles, I had to get off the treadmill to use the bathroom. But, on the fourth night, I feel asleep without any Tylenol and on the fifth day, I felt much better. The discomfort which I had braced myself for to last 2 or 3 weeks seemed to have lasted only a few days.

But, Zyprexa's clutches were not finished with me yet. Just when I thought the withdrawal period was ending, it returned with a vengeance. The next day was Saturday, time for my scheduled long run. It was hot mid-August day, about 97ºF, and humid. For hot days, I had a running route mapped out that took me past three water fountains in local parks. Re-hydrating during a long hot run was essential not only for completing the run but to avoid getting ill. I ran eleven minutes, drinking a total of about 2 quarts along the way. I didn't feel quite right after the run. At first, I thought I overdid it so I laid down for about 30 minutes after taking a shower. Following that, my wife and I went to a barbeque where I ate well and continued to drink bottled water. Just before bed, my legs felt extremely restless. I was overcome with the overwhelming need to move them. This feeling continued well into the night making it difficult to fall asleep. I also experienced the chills as well as the sweats throughout the night. Twice during the night, I got so hungry that I went downstairs to eat. The first time, I ate a bowel of cereal. The second time, I ate cookies and milk. In the middle of all this, I had a couple of bowel movements as well. I realized that I had not dehydrated myself but that I was finally beginning to feel the withdrawal side effects of Zyprexa in all of its glory.

I felt exhausted and depressed the following day but carried on just the same. I went to church and then out to lunch with my son. I ate a solid lunch which included a milk shake in order to fuel up for another long run. Later that afternoon, I went out for my run. Fortunately, it was fifteen degrees cooler than the previous day. I slowed my pace down and put my head down and just slugged through the miles one by one. To my surprise, I began to feel a little better after 7 miles into the run. I finished the nine and a half miles running a 10 minute per mile pace.

It's difficult to find any official or valid information on the withdrawal side effects of Zyprexa. Some anecdotal accounts on the Internet can be found but many believe that there are no withdrawal side effects, just the return of symptoms from the underlying disease. This was not my experience. First, I had no "underlying disease". As I indicated, I was not schizophrenic or depressed. Over the next two days, I experienced considerable GI distress, twitching, chills and sweats. My mood was down as would anyone's be who was not feeling well but I was not clinically depressed. I could still experience happiness and laugh at ordinary jokes and humor. Finally, at about 8 days after I discontinued taking Zyprexa, I began to feel well again. All of the withdrawal symptoms faded away, my mood was excellent and I could actually sleep.

I continued to stick with my training schedule and actually did a little better than I originally projected. At two weeks prior to the marathon, I took my longest training run of 18 miles. The weather remained hot and the run was difficult. I ran along the route with the water fountains in order to remain hydrated and finished the run in about 3 hours flat. This told me that I should target my marathon finish time for four and a half hours.

The day before the marathon, my wife and I traveled to Dayton. Our first stop was the Nutter Center where I picked up my registration packet. A fitness exposition in connection with the marathon was set up at the center. We looked around and visited several of the booths, picked up my commemorative tee shirt and patch and departed for the hotel. Marathoners traditionally eat pasta the night before a race, so we located a local Italian restaurant in Fairborn, Ohio, and had a nice Italian pasta dinner.

As feared, I could not sleep at all that night but it didn't bother me too much because I was used to it. I took a shower at 5 a.m. and then ate a large quantity of graham crackers that I brought with us.

The race was to begin at the Air Force Museum on the Wright Patterson Air Force Base at 7:35 a.m. We arrived at the parking lot at 6:45 a.m. It was crowded with over 3000 runners, their family members and friends. Most runners upon arrival immediately lined up at the port-o-lets, me included. From the rest area, I proceeded to the start line and located the official pacer who would run the marathon in exactly 4.5 hours. His name was Bill and he was 39. He had run 20 marathons and was very comfortable with a 4.5 hour pace. He had a global positioning system strapped to his arm and a runner's watch on his wrist. He also wore a tape around his wrist with the target split times for each mile.

I decided to run with a pacer in order to avoid running too fast for the first half of the race. Inexperienced marathon runners make this mistake often and end up walking later in the race. I wasn't totally inexperienced but it had been 22 years since I had last run a marathon and I didn't want to take any chances. About 25 people elected to run with Bill.

While waiting for the race to begin, I was nervous. Could I make it after all that I had been through? Would my brain act up late in the race when I would be pushing myself to the maximum? I knew that anything could happen after mile 20. Would my legs and joints hold up after 20 miles? Would the lack of sleep catch up with me?

I had a talk with myself. I said, "Sleep or no sleep, I will not only finish but I will not break my stride and I will not stop or walk at any point. I've done this before and I know that my training was sufficient. I will not let down the Chiari community and everyone who pledged by the mile will end up making their full donations."

I also did not want to let down Mandy. Mandy lives in New Zealand and suffers from Chiari and Syringomyelia. Several months prior to the race, Mandy posted a question on the WACMA Yahoo egroup and I responded. From that point on, we exchanged messages daily and talked about all kinds of topics. At first, we discussed Chiari and health care but over time we discussed our hobbies and interests, our families, politics, religion and many other topics. The heath care system in New Zealand leaves a lot to be desired. Patients requiring surgery often have to wait months. Mandy waited months for her surgery. Her surgery was scheduled and cancelled four separate times along the way. I "watched" her deteriorate over the months and she was aware that I was training for the marathon. Mandy finally had her surgery the day before the marathon. She would be in the ICU while I would be on the course and I did not want to let her down.

The race started on time at exactly 7:35 a.m. Our 4:30 pace group crossed the start line about one minute later. The first few miles are always a little difficult until the body loosens up and breathing gets into a rhythm. The course had two big hills, the same hill, near the start and then again towards the end. It was about a 2 degree rise over 1.7 miles. After getting past the opening hill, I established a rhythm and ran silently in the pack past the first couple of water stations. I took my first drink at the third mile and pressed on still feeling uncertain about my ability to hold up towards the end.

At about the 8-mile mark, I heard a couple of men talking just behind me so I dropped back to chat with them. We talked about technology, things like cell phones and satellite images of the world on the Internet.

Both were military career men. One turned out to be Air Force general. It was clear that he wasn't in the mood to converse, instead appearing to want to focus on his running. The other was much more talkative and was stationed at the Wright Pat base where he had been working in the same office for 40 years. He was 69 years old and his name was Allen. Allen was running his ninth marathon. He started running marathons at the age of 40 and had even run a 52-mile extreme marathon completing it in 12 hours. As we talked, I couldn't believe all the things that Allen had accomplished. Some of his accomplishments included canoeing 500 miles down the Wabash River, roller skating 400 miles across Holland, cross country skiing with 4000 Russians in the Russian Arctic, Kayaking around Iceland, and biking 4000 miles from Dayton, Ohio, to the Arctic Circle in Alaska. He had also authored ten books and planned on competing in the modified iron man competition in Hawaii for his 70th birthday.

Running with Allen was very therapeutic. It took my mind off the running and before I realized it we were approaching the 15-mile mark. At that point, Allen began to fall back and I went ahead of the pace group. I felt unusually strong through miles 16 and 17 but some discomfort began to emerge at mile 18. The strain however was not enough to slow my pace and I pressed on.

At that stage in the race, people who had gotten out ahead of me and who overestimated their ability began to walk and I began to pass numerous runners for the next several miles. At mile 22, I found myself facing the big hill on the return. I turned my head to a runner on my left who appeared to be slowing down and said, "Oh, shit!" He could see that I was determined not to break my pace and it seemed to challenge him. He stayed with me up the entire 1.7 mile hill. I don't think he liked the idea of someone who was easily 15 years his senior passing him at that point.

While climbing the hill was challenging, I found coming down the hill much more painful particularly to my quads which were aching with each and every step. I pressed on.

After descending the hill, I could see the finish area about 1 to 1.5 miles in the distance. My legs were throbbing with each step and I could feel new blisters forming on the arches of my feet. About a half mile from the finish line, the road was filled with spectators many of whom were faster runners who had already finished the race. They shouted the usual cheers, "looking good!", "keep it up", "you're almost there". The final stretch of about 300 yards took the runners between two rows of planes on display in front of the museum. As I entered the final stretch, I could see the finish line in the distance and the large timing display. A young couple who had been with me for the last five miles, sprinted ahead of me. I recalled a marathon that I had run 22 years earlier in which I decided to sprint the last 100 yards in order to beat a competitor who I had gone neck and neck with throughout the race. As I speeded up, I pulled a ham string and ended up hoppling across the finish line. Based on that experience, I decided to let the young couple go ahead of me although I may well have been able to hold them off. I hit the finish line at 4 hours 28 minutes 19 seconds. Several uniformed officers were at the finish line presenting the runners with medals. The one directly in front of me was a short lady. I bent over so that she could place the medal around my neck and nearly fell over. Another officer jumped over and assisted her in helping me stand up. With that, I thanked the Lord and waved at my wife who was standing behind the gate.

I made it, all 26.2 miles, with no sleep and 2 minutes ahead of my target finish time. My 6-year journey back to wellness was finally complete.

Chapter 12
Closing the Book on Chiari?

With the completion of the marathon it was time to close the book on Chiari. It was time to let it go. It had ruled my life long enough and disrupted nearly every system in my body and mind. It caused me to be sleepy for months. It kept me from sleeping for years. It had put me in a wheelchair. It had even checked me into the psych ward.

What had exactly happened, I was not sure. Did I have a simple case of Chiari or was it more complex than that? Did I have Chiari and clinical depression, depression brought on solely by worry? Or, did the Chiari force depression upon me directly? Perhaps I had battled both Chiari and Chronic Fatigue Syndrome simultaneously. Perhaps I really had Chronic Fatigue Syndrome and an asymptomatic Chiari malformation. Could the surgery have been unnecessary? Where did the Restless Legs Syndrome fit in? Was it independent? Was it a sequelae of Chiari? Was it a side effect of Zyprexa or one of the other medications?

The answer to these questions in my case will never be known. After all, no one knows why so many people with Chiari malformations have no symptoms until later in life. What changes occur over time that lead to the emergence of symptoms much later in life? This very basic question needs to be answered but will require a well-designed clinical longitudinal study as a first step. Who will fund such a study?

Other aspects continue to trouble me. The symptoms I experienced were very real and very severe. Why did some doctors blow me off as though I was nuts? As far as they knew, I was a model of perfect health both physically and mentally until I began complaining of symptoms in my mid forties. Why would a patient who never had health problems through almost five decades of life and who was a highly educated professional not be taken more seriously? Was there no credibility in my past track record? A generation or two ago, patients often stayed with the same doctor for life. Doctors got to know their patients and what was truly abnormal for them. Today with a more mobile society, changing health plans and physician networks, patients have become numbers to their doctors.

Why are doctors stuck on the notion that a 5mm herniation of the tonsils as viewed in a two-dimensional slice from an MRI scan is not sufficient enough to cause symptoms? Viewing a three-dimensional object in two dimensions is simply incomplete. Why isn't Cine MRI ordered on patients more often who complain of symptoms and have borderline herniations of 3 to 5 mm? Could it have anything to do with the fact that many physician groups get financial incentives from healthcare insurers for keeping diagnostic tests costs down?

Where does clinical depression fit in? Depression can cause many symptoms including pain and sleep disturbances. Symptoms of depression and anxiety are common in patients with Chiari malformations. Mueller and Oro report that nearly 50% of the 265 Chiari patients they studied presented with depression. On the other hand, People suffering from depression without Chiari malformations possess many of the same physical symptoms as those with Chiari malformations. How can this be better sorted out? The answer, to some extent, is probably Cine MRI. Patients with minimal herniations (3 to 5mm) and unexplained physical symptoms must not be written off as merely depressed. For these patients, Cine MRI should be performed to determine if CSF flow is blocked. The Chiari patient with restricted CSF flow and depression needs the treatment services of both the neurosurgeon and the psychiatrist.

Lastly, why is there a total void when it comes to recovery and rehabilitation for decompressed Chiari patients? In my case, I was sent home with instructions to straight cath myself. No home visit nurse was offered or discussed. I could hardly bend my neck enough to perform the procedure. My hospital rehabilitation evaluation went well. I could walk without being thrown off balance. I could climb steps. The conclusion was that no rehabilitation program was necessary. What about the pain in my arms every time I raised them or the nausea I would experience after using my arms to any extent? More care and consideration is given in cases of simple pinched nerves or carpal tunnel syndrome than when the grand daddy of all nerves is compressed – the brain stem.

From beginning to end, there are problems in this country when it comes to Chiari. Going back again to the pinched nerve concept, Chiari amounts to the brain stem and upper spinal cord being pinched. This is a significant neurological compression. It deserves to be treated seriously. A much better job has to be done in teaching new medical students and in updating currently practicing physicians. Relative to diagnosis, there are too many neurologists and neurosurgeons who believe that the herniation has to be greater than 10mm before symptoms can be attributed to Chiari. This is old school and just plain wrong. Herniations as small as 3 to 5mm can cause significant symptoms. Also, looking at a few two-dimensional MRI slices is often not sufficient to determine if adequate space exists for CSF to flow. Cine MRI must be used more often to confirm the diagnosis before writing off the patient as depressed or being a hypochondriac. Good follow-up care is needed and the neurologist should play a key role here. Complete evaluations should be given to patients who continue to complain of symptoms following surgery beyond several months. The work up should include MRI and Cine MRI of the head and spine.

More research is needed. What causes the malformation in the first place? Is there a genetic component? What can we learn from dogs as Chiari has also been observed in canines? What causes the emergence of symptoms often later in life? Longitudinal studies are desperately needed.

Finally, there is a movement out there, albeit small, to spread awareness and answer some of these questions. Groups like the Chiari Institute, the American Syringomyelia Alliance Project, the World Arnold Chiari Malformation Association, the Chiari & Syringomyelia

Patient Education Foundation, and others, are to be commended and supported for their fine efforts.

Appendix 1
Medical/Technical Information on Chiari

Chiari (pronounced kee-ar'-ee) Malformation Type I of the brain or CMI is an uncommon congenital birth defect.

In CMI, the cerebellum is elongated, protrudes through the opening at the base of the skull or foramen magnum, and resides in the upper spinal canal. Normally, only the spinal cord passes through the foramen magnum and down into the spinal canal. As such, sufficient space exists for the cerebral spinal fluid or CSF to drain from the skull into the spinal canal. In CMI, the additional space at the foramen magnum and in upper spinal canal is restricted by the presence of the lower cerebellum or cerebellar tonsils. Over time, processes not well understood by medical science further restrict this small but critical space essential for CSF drainage. Eventually the CSF becomes significantly blocked and compression of the hindbrain, brain stem, and lower cranial nerves occurs.

The compression results in a multitude of neurological symptoms. The constellation of symptoms varies from one individual to the next. Symptoms can emerge suddenly or very gradually. For some, the symptoms are mild. For others, the result can be devastating, crippling and even deadly in rare instances.

Some of the more common signs and symptoms are headaches, dizziness, pain, difficulty swallowing, difficulty sleeping, shortness of breath, tingling, numbness and burning of the extremities, weakness, fatigue, and nausea. Symptoms can be unilateral or bilateral. Emotional and mental problems are also frequently present. The mean age for the emergence of symptoms is about 25 but symptoms can surface in children as young as six months. On the other end, in some adults, symptoms are often delayed until the fourth or fifth decade of life.

In addition to all the problems caused by compression to the brain, many patients (about 65 to 80%) encounter a complicating condition known as Syringomyelia (pronounced sear-IN-go-my-EEL-ya). Syringomyelia (SM) is the medical term for a cyst in the spinal cord or a syrinx. According to one theory, once the CSF is blocked, it often penetrates the spinal cord to escape the skull. Like Chiari, the degree of Syringomyelia is highly variable. Also, the size of the syrinx does not necessarily correlate with the number or severity of the resulting symptoms. Some of the more common symptoms associated with this condition are pain, the inability to sense the difference between hot and cold, weakness, and bowel and bladder problems. Patients with Syringomyelia sometimes present with second-degree burns on their bodies as a result of stepping into hot showers after falsely determining that the water was cooler with their hand.

In 1883, John Cleland first described hindbrain abnormalities. The German pathologist, Hans von Chiari, acknowledged Cleland's paper and classified cerebellum alterations by types I, II, and III in 1891. Although the anatomical anomaly has been known for over a century, the syndrome remained poorly understood and accurate diagnosis hampered by the lack of sensitive and discriminating imaging technology. As a result, most doctors to the present day have never correctly diagnosed a single case. CMI is usually misdiagnosed as multiple sclerosis (MS), fibromyalgia (FM), chronic fatigue syndrome (CFS) or clinical depression. Recent research by Dr. Dan S. Heffez suggests that as many as 20 to 25% of the 4 to 5 million patients in America diagnosed with FM/CFS have been misdiagnosed and actually have CMI[1]. Further investigations are needed to verify this observation.

Patients often suffer for years before an accurate diagnosis is made and surgical treatment rendered. According to an unscientific poll taken by the World Arnold Chiari Malformation Association or WACMA, more than 75% of patients waited 5 or more years before receiving a correct diagnosis. It was not until the advent of Magnetic Resonance Imaging or MRI that accurate diagnosis even became possible. Sometime after that, it was arbitrarily decided that tonsillar herniations below the foramen magnum greater than or equal to 10 millimeters (mm) constituted Chiari. Thus, the 10mm cutoff quickly became part of the medical folklore and neurologists and neurosurgeons alike were taught that smaller herniations did not cause symptoms. More recent research by Thomas H. Milhorat and colleagues, however, has confirmed that herniations as small as 3 to 5mm can cause significant symptoms[2].

[1] DS Heffez et al, Eur. Spine J., Vol. 13, No. 6, p. 516, Oct. 2004.
[2] TH Milhorat et al, J. of Neurosurgery, Vol. 44, No. 5, p. 1005, May 1999.

Unfortunately, the medical establishment is slow to change particularly for something that is still perceived as rare. However, more and more specialists are now becoming aware of the new 3 to 5mm cutoff criterion. Importantly, an increasing use of a special form of MRI called Cine MRI is now being employed to confirm cases involving borderline herniations. Cine MRI is a dynamic scanning technique that allows the diagnostician to actually observe the flow of CFS as it leaves the skull and enters the spinal canal. It can determine exactly where the restriction of flow is occurring as well as measure the CFS flow rate across the foramen magnum. Using this technique, which was originally pioneered for heart disease, it has been determined that the CSF volume in the hindbrain is reduced in CMI patients on average by 60%.

With the exception of acetazolamide, drugs are largely ineffective in treating CMI. Acetazolamide is a potent inhibitor of fluid secretion and is typically used to treat certain types of glaucoma, edema, and epilepsy. It is used in CMI in an attempt to reduce CFS output to generally decrease hindbrain compression and relieve symptoms. Unfortunately, in many cases, acetazolamide is ineffective. Because medications are not particularly effective, the neurologist has played a minimal role in diagnosing and managing the condition. The only effective treatment for CMI, once prominent symptoms emerge, is decompression surgery. As such, neurosurgeons almost exclusively treat the condition and subsequently know the most about it. However, even within the community of neurosurgeons there remains a high variability with respect to knowledge of and experience with CMI. Thus, patients in which radiographic MRI evidence of cerebellar tonsil herniation is found need to identify and seek out the services of a neurosurgeon with considerable experience in performing decompressions of the foramen magnum.

This is different compared to other diseases and medical conditions. For example, with heart disease, it is the cardiologist who typically makes the diagnosis. If surgery is indicated, then the patient is referred to the cardiac surgeon. After surgery, the patient returns to the cardiologist for the appropriate follow-up care. However, in CMI, the patient needs to take the unusual step of first going to the surgeon for both diagnosis and treatment. Unfortunately, after surgery, the Chiari patient is pretty much left on his own to recover which is a major problem in many cases.

The surgical treatment of CMI is nothing less than a modern miracle. The basic idea is to decompress the brainstem by removing bone and expanding the tough elastic membrane that surrounds the brain and spinal cord known as the dura. This is achieved by first removing a relatively small amount of the skull in the back of the head just above the spinal column. Next, the posterior arch of the first and sometimes the second vertebrae is removed. After the removal of the restricting bone, the dura is opened to expose the tonsils. Because the tonsils are misplaced in CMI and subjected to the motion of head turning, it is hypothesized that the body responds by forming adhesions between the tonsils and surrounding tissue, like the upper spinal cord, in an attempt to stabilize the lower cerebellum. These adhesions contribute to the restriction of space needed for normal CSF flow. The adhesions are then carefully removed by the neurosurgeon with the aid of an operating microscope. Because the tonsils have no known function some surgeons remove them completely but this step remains somewhat controversial. After the removal of adhesions and/or the tonsils the dura is then expanded with a graft. Traditionally, the graft material used has been bovine pericardium tissue that has been specially processed. This material, however, is now giving way to other materials and sealants such as DuraSeal®, a patented synthetic, absorbable hydrogel, have been recently approved by the Food and Drug Administration.

After duraplasty, the incision is closed and normal CSF flow returns. The goal of surgical decompression is to arrest the progression of symptoms, however the reversal of symptoms is often achieved to various degrees, particularly in cases without SM.

CMI has no racial, social or economic barriers. Chiarians differ in terms of their symptoms and the impact of their symptoms on their lives. Complicating the picture for many is the psychological component of the syndrome. In CMI, the brain is under abnormal compression. It only stands to reason that the depression, anxiety, obsessive compulsiveness, and panic attacks that are so commonly observed in Chiari are possibly the products of neurotransmitter imbalance due to pressure differentials. Yet, the relationship between brain compression/decompression and psychological disease is probably the least researched aspect of CMI. As a result, hundreds if not thousands continue to struggle after decompression surgery.

CMI is not rare. It is more correctly classified as uncommon. In a recently published review article by Marcy C. Speer[1], the prevalence of Chiari is estimated to be somewhere between 1/1200 and 1/1500 and published data by Ajay K. Bindal[2] suggests that the percent of symptomatic cases is about 80% although the number of patients studied was small. Perhaps more provocative is the recent research of Dr. Heffez cited earlier, which implies that as many as one million symptomatic Chiarians may exist in America.

Chiarians usually do not appear sick yet this couldn't be further from the truth. But, family members of those afflicted often do not recognize this, particularly when attending physicians incorrectly subscribe complaints to other aliments perceived as less serious. Family members must come to realize that the complaints of their loved ones are real and deserving of proper medical intervention. The additional appendices in this book provide detailed information on various aspects of CMI such as clinical signs and symptoms, related survey data and resource organizations.

After surgery, there is a strong tendency for the patient to remain in a shell and to avoid doing things that, prior to surgery, exacerbated symptoms.

[1] MC Speer et al, J. Genet. Couns., Vol. 12, No. 4, p. 297, Aug. 2003.

[2] AK Bindal et al, Neurosurg., ol. 37, No. 6, p. 1069, Dec. 1995.

This is understandable as many with Chiari have lived for extended periods of time where just engaging in simple day-to-day activities like walking, gardening, or mowing the lawn brought on nausea, vomiting, dizziness, headaches, breathing problems or heart palpitations. This book is also about bravery. The recovery process requires taking some risks or testing the waters so to speak. By taking small risks at first and carefully listening to the feedback from the body, the decompressed Chiari patient can begin to move forward in the recovery process. There will be setbacks along the way but the determination to move forward must dominate the mindset. One must not expect to recover in just a matter of weeks or months after surgery.

Appendix 2
Signs and Symptoms of Chiari

Sign/Symptom	Percent of Patients Studied	
	Milhorat et al[1]	Mueller & Oro[2]
Headache	nr[3]	98
Difficulty sleeping	nr	72
Fatigue	nr	59
Eye pain or pressure	63	nr
Floaters/flashing lights/distortions	55	nr
Blurred vision	48	57
Sensitivity to light	28	21
Double vision	20	15
Loss of peripheral vision	7	nr
Dizziness	57	84
Disequilibrium	51	46
Pressure in ears	46	nr
Ringing in the ears	38	56
Hearing loss	36	16
Vertigo	20	6
Oscillating viewed images	13	nr
Sensitivity to sound	7	nr
Difficulty swallowing	43	54
Sleep Apnea	38	nr
Hoarseness	31	41
Tremors	26	nr
Palpitations	26	2
Poor coordination	23	nr
Throat pain	16	nr
Facial pain/numbness	16	32
Fainting	13	nr
Shortness of Breath	9	57
Hypertension	8	2
Hypotension	nr	2
Numbness/tingling	59	62
Insensitivity to pain	40	nr
Burning sensation	29	nr
Poor position sense	23	nr

Impaired temperature sensation	22	nr
Muscular weakness	57	69
Neck pain	nr	67
Spasticity	30	nr
Muscular atrophy	9	nr
Nausea	nr	58
Vomiting	nr	15
Trophic phenomena	20	nr
Urinary incontinence	17	nr
Impotence	24	nr
Fecal incontinence	2	nr
Depression	nr	47
Anxiety	nr	30
Memory problems	nr	45
Involuntary eye movement	26	5

[1] TH Milhorat et al, J. of Neurosurgery, Vol. 44, No. 5, p. 1005, May 1999.

[2] DM Mueller and JJ Oro, J. Am. Acad. Nurs. Prac., Vol. 16, Issue 3, p. 134, March 2004.

[3] nr = not reported

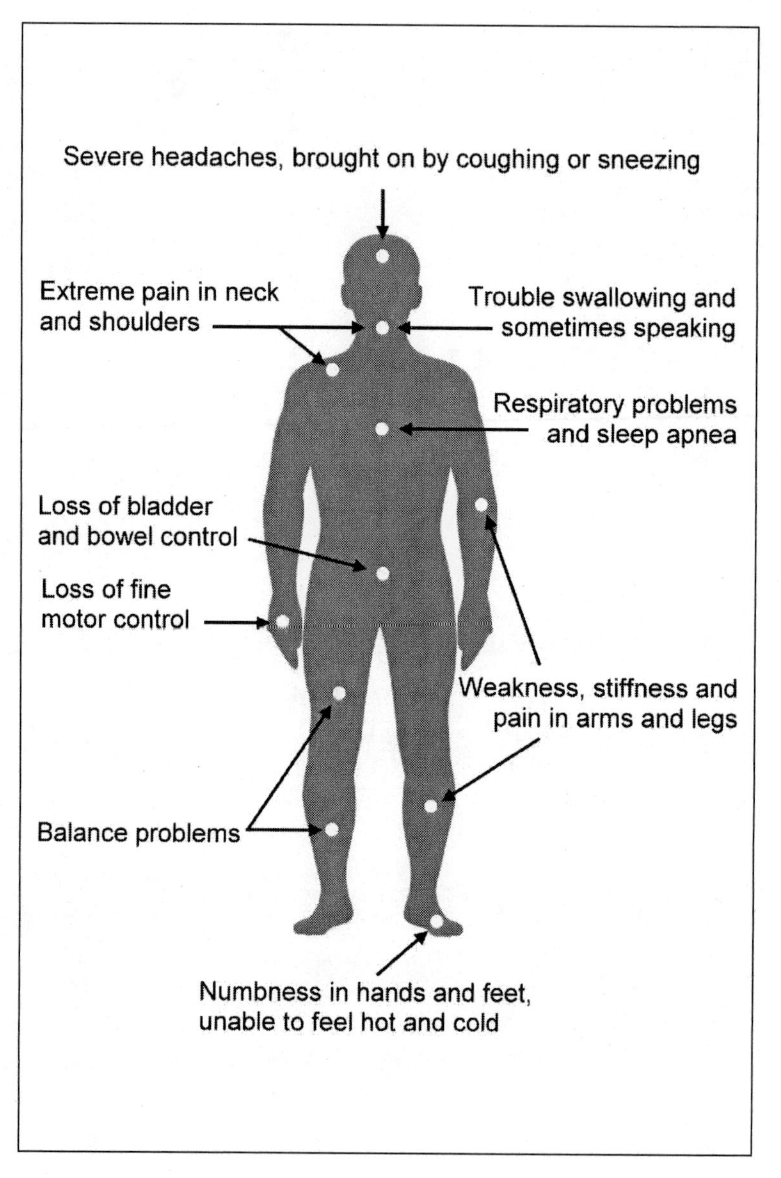

Illustration courtesy of Conquer Chiari

Appendix 3

Resources

Conquer Chiari or Chiari & Syringomyelia Patient
Education Foundation
320 Osprey Court
Wexford, PA 15090
http://conquerchiari.org/index.htm

World Arnold Chiari Malformation Association
http://www.pressenter.com/~wacma/

American Syringomyelia Alliance Project or ASAP
P.O. Box 1586
Longview, Texas 75606-1586
http://www.asap.org/

The Chiari Institute
865 Northern Blvd.
Great Neck, NY 11021
http://www.chiariinstitute.com/

National Institute of Neurological Disorders and Stroke
http://www.ninds.nih.gov/disorders/chiari/chiari.htm

University of Washington Chiari Malformation Clinic
http://depts.washington.edu/neurosur/chiari/

Appendix 4
Selected WACMA Survey Data

(These data were not collected scientifically nor can they be verified)

How many years from symptom onset to diagnosis?

Years	Percent of Responders
1	21
2	11
3	8
4	6
5 or more	54

Number of responders = 135

Age at onset of symptoms

Age	Percent of Responders
0-10	13
10-20	18
20-30	30
30-40	28
40 and over	11

Number of responders = 152

Length of herniation

Length	Percent of Responders
0-5mm	18
5-10mm	31
10-15	32
15-20	9
Over 20mm	9

Number of responders = 116

Sex

Sex	Percent of Responders
Male	8
Female	91

Number of responders = 145

What were you misdiagnosed with before learning you had a Chiari malformation?

Diagnosis	Percent of Responders
Multiple Sclerosis	17
Depression	35
Lymes Disease	1
Chronic Fatigue Syndrome	13
Fibromyalgia	12
Carpal Tunnel Syndrome	12
Menieres Disease	8

Number of responders = 88

Are you glad you had the surgery?

Response	Percent of Responders
Yes	93
No	7

Number of responders = 89